Alternative Crops for Drylands

Advance praise for *Alternative Crops for Drylands*:

"*Alternative Crops for Drylands* profiles numerous useful plants from the world's deserts and arid lands, from the African savannahs to Australia and the Chilean Atacama. It features most of the crops I feel are most promising for drylands as well as a great many that were new to me. This book is an important and unique resource for the productive restoration of arid lands. In addition to covering established crops it points out over a hundred more species worthy of domestication. Gardeners, farmers, permaculturists, and aid workers in arid climates would do well to spend some time with this important manual."
-**Eric Toensmeier**, Author of *Perennial Vegetables*

"*Alternative Crops for Drylands* is bound to be a classic for people living and working in brittle environments. As a Permaculture Consultant for dryland environments in the Americas, Middle East and in Africa, I will refer to this well organized book often as a key resource in helping establish resilient and stable food systems. Thank you Scott O'Bar for your research and compilation of this important book for our day and for generations to come."
-**Warren Brush**, Instructor of Permaculture Design for International Development Professionals and Founder of Quail Springs Permaculture

"Coming at the book from an agricultural development perspective, I am encouraged to see that it includes a host of multi-purpose, underutilized trees and other plants suited to dryland regions of the tropics. The detailed crop descriptions, high-quality photos, and literature sources listed for each species all contribute to making this book a tremendous research tool for practitioners seeking ways to improve livelihoods of resource-poor farmers."
-**Tim Motis, PhD**, Agriculture Research Director, ECHO, Inc.

Alternative Crops for Drylands

Proactively Adapting to Climate Change and Water Shortages

Scott O'Bar

∞
Amaigabe Press
Santa Barbara, California

Alternative Crops for Drylands
Proactively Adapting to Climate Change and Water Shortages
by Scott O'Bar

Amaigabe Press
915 Randolph Rd.
Santa Barbara, CA 93111 U.S.A.
amaigabe@yahoo.com

Edition ISBNs
Softcover: 978-0-9882822-0-9
PDF: 978-0-9882822-1-6

Library of Congress Control Number: 2012916698

First Edition 2013
Printed in China

Cover Photo: Etan J. Tal
Pictured: *Cereus repandus* fruiting in the Negev Desert of Israel
Cover Design: Mark Viergutz
About The Author Photo: Mark Viergutz

Note To The Reader:
The information contained within this book is intended as a review of the literature, a presentation of existing information and that derived from the author's firsthand experience, and is not intended to be exact or true in all cases/circumstances. Furthermore, the author and publisher are not responsible for any ill effects experienced through the consumption and/or misidentification of any of the plants and their parts mentioned in this book. In all cases, it is your responsibility to do your own research and to determine the edibility of any plant before making any decision to consume the plant or any of its parts and/or byproducts.

Special Thanks To:
Mark Viergutz
Trinity Lila
Jerry Sortomme
Matthew B. Johnson
Sue Reinhart
Mike Kauffman
Russ & Catherine Salzgeber
Frank Tozer
Elizabeth Robinson
Tonya Schultz
Stephen Tredwell
My parents

Contents

List of Figures

Foreword

Alternative Crops for Drylands: Proactively Adapting to Climate Change and Water Shortages is an impressive portrait of edible and useful plants that Scott O'Bar has selected from Earth's wayward arid, desert and water-stubborn environs.

Plant life is tenacious and life forms can be found in the most peculiar and adverse spots on this planet. From the depths of the continental land mass, in steaming, bubbling thermal pools, or camouflaged as skid mark smudges within ancient polar ice, or floating, parachuting and in orbit around the upper atmosphere of our planet, or hiding in the deepest ocean trenches, scurrying around smoking, volcanic vents – life clings, hides and thrives just about everywhere it has any chance to exist, and, until recent times, who knew just how grand this diverse story is?

These are the bookend places of the biosphere...of the third planet in this solar system. Ask Hubble. Earth is a rare and unique water world. The most curious of living creatures are being revealed to humankind as we discover our worldly perimeters. As we delve into almost every nook and cranny of Earth, extremophiles are revealing themselves. Yet as tenacious as life is, life's personality and character is distinctly fragile as it encounters its own boundary edges...this is the paradox, the tapestry of limits.

Scott has chosen to explore those plant forms and plant niches that are defined as... water-limited. DROUGHT...it is one of those words that signals to us collectively... be careful, beware, be ready. Drought means to our society that there will be less water than normal. In the face of extreme drought, severe restrictions, increasingly life-altering consequences, and perhaps worse conditions are looming.

During the summer of 2012 the United States endured a near worst case scenario of...very severe drought. The negative domino effects of this 2012 drought, were arguably as entrenched and widespread as those experienced in the dangerous Great Dust Bowl disaster of the

20th century. Wise decisions and future planning got us through that natural, American Dust Bowl disaster.

Now we are coming face-to-face with not only an episode of drought, but Mother Earth is displaying to us a pivotal chapter of worldwide climate change, and indeed evidence of global warming that is rapidly defrosting and melting a majority of the glaciers and expanses of ice caps worldwide. The new norm is predicted to be extreme weather conditions and changing climate patterns. It seems that Gaia is making adjustments. So, it would be wise to heed the signs and get ready to counter. Plant selection is a way to make adjustments.

The California Rare Fruit Growers (CRFG), to whom I am a lifetime member, is a group of horticulture enthusiasts who grow a gamut of edible crops from all points on the compass. They have a botanical arsenal of plant choices. Scott and groups like CRFG can collectively assist us to logically maneuver our way through some would-be horticultural perplexities. This is a story of survival...we have options...and some quite delicious and tasty options.

Science has given us the advantage of seeing the bigger picture, and history has given us the advantage to observe those successful people and their ideas that have influenced long-term sustainable decisions and indelible solutions. There are 7+ billion humans alive on Earth as Scott releases this book. The 8 billion population threshold will be reached during most of our lifetimes. Where will the newcomers live and expand out to, in this world? What will our various lifestyles adapt toward? What degrees of quality of lifestyle shall exist? There are choices and options. Water matters. Water is key.

Water is an absolute finite natural resource, and necessary for every living thing. Quality fresh water is more precious than any resource we must manage. For humans to have good quality and sustainable lives into the next generations, we must, above all, orchestrate our worldwide water consumption. Corporations are buying up the water rights of almost any water source or watershed on Earth. Those who control the water shall have the ultimate control. Sharing and distribution of water is crucial.

Scott, in this book, brings to our attention a possible approach to fashioning a plethora of wise horticultural options and choices. He has indexed an interesting and important listing of plant genera & species that can be problem-solving plant choices. Cultural and habitat information is included.

Generally, lower mandatory water allotments are in our future. Yet, things are do-able and can be pleasant.

In arid and dry climates, primitive and enduring cultures absolutely had to deal with the wise management of water for drinking & cooking, for agriculture & manufacturing and in sustaining & nurturing the "greater environment." History records that those individuals and entire cultures that did not strike a balance and equilibrium between human and non-human, between organic and inorganic abruptly or miserably withered away. Ecology is the study of this balancing act. Many people believe our future will not only impact individuals and cultures, but that our future will also have a larger impact on the planet. Nonetheless, we can each finesse our apt approach for heading into the future.

From 1980 to 2002 I was the Department Chair and the only full-time instructor of the Santa Barbara City College Environmental Horticulture department. In that capacity I designed the two career-horticulture outdoor classroom laboratory gardens; 1) The Santa Barbara City College - Lifescape (a worldwide edibles garden) and 2) The Chumash Ethnobotanical Preserve (a locally indigenous botanical display). These adjacent gardens are where community college practical "laboratory" classes in horticulture take place.

During my 22 year time frame of teaching Environmental Horticulture, several thousand students took higher level and career level horticulture classes in the Environmental Horticulture program. No two individual student-apprentices ended up in the same place...A different result at the end of each journey. Horticulture classes allow each student to gather facts and carve a path toward their future and goals.

Scott graduated from the Environmental Horticulture (EH) Department at Santa Barbara City College (SBCC) with an EH Associate Degree of Science. Scott's vision of this book is to empower those who want to be ahead of the tide – to be problem solvers and pathfinders in a dry future. Scott is steering his future course. It is an admirable journey, and an important venture.

There are so many plants Scott has highlighted in this book. Here are just a few examples: Acacia, Baobab, Num-Num (Carissa), Myrtle, Opuntia cactus, selections of palms, capers and many more. He has selected dryland edible plants from the Northern Hemisphere and

Southern Hemisphere – from flora environs that include: coastal & inland deserts, savannahs, worldwide Mediterranean zones, chaparral & sagebrush niches, etc. There is something for everybody from everywhere to delve into. Some plants you will recognize, most you will be introduced to for the first time.

In retirement, I am the Gardens manager and consultant of a horticultural project called, La Huerta Historical Gardens at the Santa Barbara Mission in Santa Barbara, California. Scott is a volunteer participant in this one and a half acre, Huerta orchard-garden – a living museum component of Old Mission Santa Barbara. This garden is a collection of documented plants (1769-1834) that the Spanish introduced and cultivated in Alta California, at the 21 Franciscan Mission chain from San Diego to San Francisco. Scott and the other volunteer-gardeners install and care for a plethora of introduced, domesticated and cultivated plants of the Spanish Empire in Upper California.

Most of the California missions were situated in semi-arid, drought-prone climate zones. Back in those times the world was growing a global array of plant materials that were rugged, time tested and highly sustainable – many still surviving in the dryland environment for 200+ years. With skill, cunning, hard work, and using the art and science of centuries of agriculture & horticulture savvy, Colonial California is an example of successfully managing a dry environment.

We learn from the past, we identify the wisdom of former times and we blend in the augmented knowledge from yesterday and today. Now we head into the future, hopefully quite prepared to not only survive, but also to thrive.

The Horticulture Equation is a guide: the right PLANT in the right PLACE in the right SOIL with the right SPACING with the right CARE done at the right TIME.

Alternative Crops for Drylands will help the reader solve the horticulture equation, and can be used by hobbyists & enthusiasts, by those seriously wanting to expand the planting possibilities of farming arid croplands, and for futurists & visionaries who want to be part of the solution and not part of the problem.

Jerry Sortomme
Professor Emeritus of Environmental Horticulture
Santa Barbara City College

Introduction

Witness desertification, soil erosion, soil salinization, crop failure, atmospheric moisture failing to precipitate over the land, dehydration, malnutrition, famine. The available fresh water is diminishing, and global demand for that water is rising.[1] Already 1.1 billion people lack access to safe drinking water, and more water shortages are predicted worldwide.[2] The world is in need of solutions in order to avoid water shortages, which could lead to widespread famine. The earth contains a plethora of resources (most of which are neglected), and can provide for us. However, these provisions come with strings attached. The earth has a set of rules, which, if bent, have disastrous repercussions. In many cases, modern-day culture is based on a period of borrowed time in which these repercussions have not yet fully manifested themselves. We are in debt to the earth, and this book will help us learn how to play by the rules, not simply as a way to get out of debt, but as a means to ensure that we thrive as a species into the indefinite future.

One manifestation of an unsustainable culture is shown by our water use. The majority of the world's use of freshwater (70%) is allocated to irrigation.[3] Much of this irrigation water comes from ancient aquifers that are being depleted faster than they are able to replenish themselves. Overpumping of water in China has caused the water table to drop at a rate of one meter per year,[4] and in North America the Ogallala aquifer is being drained at a rate of 800 gallons (3,028 liters) per minute,[5] which has contributed to the average depth of the aquifer falling from 240 feet to only 80 feet deep.[6] Growing common, thirsty crop plants contributes to this problem. For instance, it takes 40 liters of water to produce a single slice of bread.[7]

Many environmental analyses insinuate that human population size is the main cause of water shortages, famines, and other similar problems. Overpopulation is not the issue. The problem is the misallocation of resources and knowledge. Before we decide to blame human population size for the starvation and malnutrition in the world, we should consider the following. First, much land is being diverted away from the production of food towards the production of other goods such as drugs for the black market. Also, governments subsidize growing crops, such as corn, for ethanol production, which yields less energy than it uses.[8] Even if ethanol production was net energy efficient, it reduces the amount of farmland available for food production.

Then we have the issue of ornamental landscaping. In many affluent areas freshwater is being used to maintain ornamental landscapes, which give no yield other than aesthetics. In the U.S. for example, there is more acreage planted to lawns than to the eight most irrigated crops combined.[9] Also, in affluent areas, people house pets, which consume human-quality food and ingredients. More importantly, perhaps, is the reality that a large portion of the human food found in supermarkets goes to waste.

Finally, turning to conventional agriculture, we witness a myopic view towards the land, even though the land requires dynamic, long-term stewardship. A monoculture is the opposite of the diversity found in any given natural ecosystem. Therefore, monocultures require endless energy inputs to maintain their productivity. Conventional agriculture is successful at producing the most amount of food from a given area of land in the immediate future, but it ruins the land for future generations. My generation will be the generation to face the consequences of a global society based on decades of conventional, industrial agriculture. We need food production systems that regenerate the land.

There are many technologies that could help humanity, which are being suppressed and/or not used to alleviate the problems we face. One such example is rainmaking technology, which has the ability to artificially generate rainstorms in otherwise dry locations.[10] However, rather than brainstorm all the technological innovations, potential remedies, and other ideas that symptomatically treat the manifestations of our fundamental problems, let's acknowledge the root of the problem. It is more efficient to discover the cause of an illness than the treatments designed to combat the illness' symptoms.

In the context of growing food for the inhabitants of dry regions, most problems arise simply from one break from logic: planting plants where they are not supposed to grow. The ancestral ethnobotanical knowledge of many regions has been ignored, and subsequent generations have regarded their useful plants as "weeds" while simultaneously becoming dependent on archaic models of aid, which involve shipments of conventionally-grown, nutrient-deficient, industrial food. In North Africa, for example, many people have grown accustomed to the colonial influence of eating processed couscous, white bread, and white rice. Such "staple" foods are often much less nutritious when compared to the edible, wild plants that are adapted to growing in North Africa.[11] Given that climate change has been shown to exacerbate desertification in many parts of the world[12], it is imperative to establish multi-species, drought-tolerant food production systems, which enrich the land while producing food.

Although nearly all the planetary bodies in our solar system are undergoing climate change,[13] many people who are either misled and/or working for carbon trading/carbon tax schemes, have unceasingly preached about our peril lest we decrease the amount of carbon dioxide released into earth's atmosphere. While I feel inclined to admit that the greenhouse effect is more than just a theory, humanity's production of greenhouse gases is not the principal driver of climate change, but is rather more akin to the effect that the emission of flatulence has in the middle of a wildfire.

In any case, reforesting and re-greening vast stretches of deserts and desertified land has the proven effect of harnessing carbon dioxide and converting it into oxygen through a miraculous process known as photosynthesis. Furthermore, vegetation eventually creates soil rich in organic matter, which has the ability to sequester carbon dioxide.

Growing edible plants in harmony with the earth, and using natural ecosystems as a design model allows for the sustained presence of "food forests" on this planet. These systems require few additional inputs once established, and may remain in balance for years even without the help of humans. The plants featured in this book should be used to construct artificial ecosystems, which achieve the same balance and permanence as natural ecosystems. Successful implementation of such systems means there would be a sustained generation of local resources, which would lead to a decreased dependence on foreign aid, an increased social welfare with more leisure, more time for higher pursuits, greater physical and emotional health, and an overall sense of satisfaction.

Most of the species presented in this book are trees, and most of these trees have multiple uses. Trees are able to grow deeper roots than common crop plants, and are thus more likely to survive prolonged droughts. Furthermore, trees are fairly low maintenance, and give proportionately large yields for many years. Many people do not consider planting tree crops simply because they do not want to wait a few years for their plants to mature and start giving yields. Another big emphasis in this book is on staple crops. Achieving a diet based on drought-tolerant staple crops will be an important feat in the near future.

To predict some criticism of this work, a potential drawback of many of the species in this book is that they are not very well developed from a breeding perspective. Many of the species are still "wild" plants, which lack the phenotypic selections necessary to generate a new "crop" plant. Nevertheless, the four plants that now make up two thirds of the average human's caloric intake[14] were wild plants at one time. Indeed, all the common, modern-day crop plants were once wild, unimproved plants. Many of the plants featured in this book, already have the advantage of good taste – even as a wild plant. With just a little breeding work, these plants may rise to culinary prominence in many dry regions of the world.

In most cases, the plants featured in this book are suitable for cultivation in areas that receive less than 500mm (20in) of average annual rainfall, and many of the plants are able to grow successfully in areas with less than 100mm (4in) of average annual rainfall. The featured plants represent a variety of dryland climates – from Mediterranean climates, to temperate deserts, to hot deserts, to semi-arid tropical savannas. This book is meant to be used as a handy reference guide for aid workers, and for people interested in conserving water even while producing their own food. This book is designed to arm its readers with a botanical knowledge for eradicating one of the most impending global problems we face: drought-induced famine.

Scott O'Bar
Santa Barbara, California

Hardiness Zones

The zone information provided throughout this book corresponds to the USDA hardiness zones. These zones reflect the average annual minimum (lowest) temperature of a region. However, the zone information provided in this book generally represents the lowest temperature that a plant will survive. Nonetheless, many factors contribute to a plant's susceptibility to cold temperatures. Depending on the age of the plant, the microclimate, the duration of cold, and the humidity, a plant's cold tolerance may deviate from the zone mentioned.

USDA Zone Information:

Z1= Below -50 F (Below -45.6 C)
Z2a = -50 to -45 F (-42.8 to -45.5 C)
Z2b = -45 to -40 F (-40.0 to -42.7 C)
Z3a = -40 to -35 F (-37.3 to -39.9 C)
Z3b = -35 to -30 F (-34.5 to -37.2 C)
Z4a = -30 to -25 F (-31.7 to -34.4 C)
Z5a = -20 to -15 F (-26.2 to -28.8 C)
Z5b = -15 to -10 F (-23.4 to -26.1 C)
Z6a = -10 to -5 F (-20.6 to -23.3 C)
Z6b = -5 to 0 F (-17.8 to -20.5 C)
Z7a = 0 to 5 F (-15.0 to -17.7 C)
Z7b = 5 to 10 F (-12.3 to -14.9 C)
Z8a = 10 to 15 F (-9.5 to -12.2 C)
Z8b = 15 to 20 F (-6.7 to -9.4 C)
Z9a = 20 to 25 F (-3.9 to -6.6 C)
Z9b = 25 to 30 F (-1.2 to -3.8 C)
Z10a = 30 to 35 F (1.6 to -1.1 C)
Z10b = 35 to 40 F (4.4 to 1.7 C)
Z11 = above 40 F (above 4.5 C)

Inches - Millimeters Conversion Table

Inches:	Millimeters:
1	25.4mm
2	50.8mm
3	76.2mm
4	101.6mm
5	127mm
6	152.4mm
7	177.8mm
8	203.2mm
9	228.6mm
10	254mm
11	279.4mm
12	304.8mm
13	330.2mm
14	355.6mm
15	381mm
16	406.4mm
17	431.8mm
18	457.2mm
19	482.6mm
20	508mm

Part 1: Nitrogen-Fixing Plants and Legumes

Legumes (members of the pea family) are special, because the vast majority of species have the ability to convert atmospheric nitrogen into a useable form, which comprises the basic elemental building block of all plant life. Plants accomplish nitrogen-fixation via a symbiotic relationship with root-infecting bacteria. These root bacteria, often of the genera *Rhizobium* and *Bradyrhizobium* convert atmospheric nitrogen for the plants, which in turn, reward the bacteria with sugars and starches.

There are few other plants that have such bacteriological advantages in the plant kingdom. Nonetheless, certain genera outside the pea family contain nitrogen-fixing species. These plants, known as actinorhizal plants, form a symbiotic relationship with *Frankia* bacteria, and are also able to exploit atmospheric nitrogen.

Nitrogen-fixing plants are useful for the sustainability of any crop production system because they are able to continuously supply a source of nitrogen external from the soil nutrient profile. Thus, the planting of nitrogen-fixing species often contributes to a net gain of available soil nitrogen. Therefore, nitrogen-fixing species are often planted in order to replenish impoverished soils, and as companion plants for other nutrient-demanding plant species.

Elegant Wattle

Scientific name: *Acacia victoriae*
Family: Fabaceae
Synonyms: Gundabluie, Bardi Bush, Bramble Wattle
Zone: 9a

Overview:

The Elegant Wattle is among a group of Acacia species, which have received research attention for their ability to help resolve famine in arid Africa. The seeds provide a nice staple crop, and can be ground into protein-rich flour. The Elegant Wattle forms a small tree reaching a height of 5-6 meters (16-20 feet). The tree spreads horizontally roughly the same distance as its vertical growth. Flowers are yellow. Branches are covered in small spines.

Figure 1: *A. victoriae* seed pods. Tucson, Arizona

Habitat and Requirements:

The Elegant Wattle is native to a wide range of Australian habitats from Mediterranean to desert, to subtropical. Throughout its native range, the species experiences a mean annual rainfall of between 100-1000mm. The tree grows well in alkaline clay soils, yet also is found growing in rocky and sandy soils. It tolerates moderate salinity. Elegant Wattle performs best in full sun.

Uses:

Fruit and Seeds:
The seeds may be eaten raw or roasted. The roasted seeds have a slight hazelnut-like flavor. Traditionally, the full-grown, green pods are also eaten by roasting over an open fire. Seeds can be ground into flour and pressed into cakes with the addition of water.

Flowers:
Like most Acacia species, the flowers provide an excellent source of pollen for bees.

Leaves:
The leaves may be used as an animal fodder.

Wood:
The wood is dense, and can be utilized as fuel or made into charcoal.

Bark:
Like many Acacia species, the bark (and also the leaves) contains the alkaloids 5-methoxy-dimethyltryptamine as well as N,N-Dimethyltryptamine, two very potent psychedelic compounds ubiquitously found throughout the plant kingdom.

Miscellaneous uses:
Elegant Wattle trees coppice and pollard readily. The trimmed vegetation may be fed to livestock or used as mulch. Closely planted trees form an effective windbreak. Elegant Wattle is also a nitrogen-fixing tree, forming a symbiosis with *Rhizobium* bacteria, which helps trees improve the soil quality.

Propagation:

Germination can be enhanced using the boiling method. Heat water until it boils, then pour the water over the seeds. Let the water cool overnight while the seeds soak. If fresh, most seeds will sprout in one week.

Potential Drawbacks:

Elegant Wattle plants will often send out root suckers. Heavy pruning or other trauma such as fire may increase the likelihood of root suckering.

Nutrition:

Seeds:	
Protein	17%
Carbohydrates	41%
Fiber	29%
Fat	3%

Source: Bushfood Australia

Related Species:

Acacia victoriae has two subspecies: *arida* and *fasciaria*. Related species include: *Acacia elachantha*, *A. tumida*, *A. thomsonii*, *A. colei*, and *A. murrayana*. The most economically viable of these are probably *A. colei* and *A. murrayana*. The former inhabits the arid and semi-arid regions of northern Australia, whereas the latter is found across central Australia. All of these species can be utilized in a similar manner as *A. victoriae*.

Unrelated Species:

Although there is no reference to edibility in any of the literature, *Acacia drummondii* is worth mentioning here because it can be grown beneath forest canopies. This shade-tolerant plant is well adapted to Mediterranean climates, small in size and nitrogen-fixing, making it an extremely useful companion plant of larger shrubs and trees.

Rooibos

Scientific name: *Aspalathus linearis*
Family: Fabaceae
Zone: 8b

Figure 2: *A. linearis*. Namaqualand, South Africa.
Credit: Arno Bredenkamp

Overview:

Rooibos is a small plant similar in appearance to broom plants (*Spartium spp.*) with needle-like leaves, which are used as an antioxidant-rich tea. The shrubs may grow up to 2m (6.5ft) tall with an equal spread, and have yellow flowers appearing in late spring, followed by hook-shaped pods that contain a couple seeds. The plants are tap rooted, and have roots that can reach 2m (6.5ft) in depth or more. Natural populations of Rooibos are threatened due to overgrazing and overharvesting of wild Rooibos.

Habitat and Requirements:

Rooibos is native to southwest South Africa, mainly to the Fynbos ecosystem in regions that experience an annual average rainfall between 300-350mm. This is a Mediterranean climate, which means that rain falls during the winter months, and there is usually a long dry season starting during mid spring and ending towards the end of fall. It is not unusual for this region to experience dry seasons lasting up to nine months, during which no rain falls.

Rooibos plants are well adapted to this dry climate and prefer to grow in sandy, nutrient-poor soil. They belong to the pea family and are thus well-adapted colonizer plants in areas where land has been cleared or otherwise disturbed. Aside from requiring full sun, in order for Rooibos to grow well, the soil must be very acidic. Rooibos naturally occurs in areas where soil pH ranges from 4-5.

Uses:

Leaves:

The leaves are made into a tea with many health benefits. Rooibos tea has become very popular in recent years on the world market.

Miscellaneous uses:

Rooibos fixes atmospheric nitrogen through a symbiosis with bacteria. Therefore, Rooibos may be used as a pioneer plant to add nitrogen to depleted soils.

Propagation:

Propagation can be done with root cuttings or seeds. Seeds benefit from a very shallow planting depth, following the old rule-of-thumb that the planting depth should be the same as the size of the seed. Germination rates can be increased via smoke treatment, scarification, and gibberellic acid treatment.

Potential Drawbacks:

Rooibos cultivation outside of South Africa has either not been well-documented or executed successfully. Rooibos plants require good drainage and a very low soil pH. In order to cultivate and propagate Rooibos plants successfully in many areas copious amounts of sand and pine bark fines should be combined to increase percolation rates while driving down pH levels to near 4.5.

Aside from its low pH requirements, the lack of successful Rooibos cultivation outside of South Africa might be due to soil-specific nitrogen-fixing bacteria and/or mycorrhizal fungi absent in non-native soils. More investigations need to be carried out involving Rooibos' relationship(s) with symbiotic soil bacteria and/or mycorrhizal fungi.

Nutrition:

Rooibos extracts have been shown to have anti-cancer effects both internally and externally (topically). Extracts have also been shown to have anti-HIV effects, and immunomodulating effects. Rooibos tea helps maintain brain health by preventing an excess accumulation of lipid peroxides in the brain, and has been shown to have anti-mutagenic, and antioxidant properties. It should be noted that the anti-mutagenic and antioxidant properties are much greater with the green form of the tea (the fresh, unfermented form).

Pigeon Pea

Scientific name: *Cajanus cajan*
Family: Fabaceae
Synonyms: Congo Pea, Gungo Pea, No Eye Pea, Red Gram, Gandul, Dhal
Zone: 9b

Figure 3: *C. cajan*. Panama.
Credit: Spencer Woodard

Overview:

The Pigeon Pea is a fairly well-known and well-researched crop in the semi-arid tropics and subtropics of the world. It is a nutritionally-rich staple crop with a well-balanced protein. The Pigeon Pea is a short-lived (3-5 years), woody, perennial shrub growing up to 4m (13ft) tall, but is usually cultivated as an annual because pod production declines after the first year. The shrubs develop a deep taproot reaching down about 2m (6.5ft). Depending on the variety, multi-colored flowers are produced 100-430 days after germination and are followed by green pods, which eventually turn brown when ripe, and contain between 2-9 protein-rich seeds. The pods do not split open while on the plant.

Habitat and Requirements:

Pigeon Pea is thought to be native to India. However, it has also been cultivated in Africa since prehistoric times. Nonetheless, no one has discovered a truly wild population of *Cajanus cajan*. In modern times the plant is commonly grown throughout the semi-arid tropics and subtropics.

Pigeon Pea performs best in areas that receive between 600-1000mm of annual rainfall, and needs conditions to be drier when flowering. Ideally, most rain will fall during the vegetative stage of Pigeon Pea's growth. Although this is the optimal range of annual precipitation, Pigeon Pea is reported to grow in areas with up to 4000mm annual rainfall, and also in areas that receive as little as 250mm of annual rainfall. In areas that receive such low rainfall, plants require ideal soil, which is a deep, well-draining, clay loam.

Pigeon Pea tolerates a wide range of soil conditions from sand to heavy clay loams and has a wide pH tolerance. Ideally, however, Pigeon Pea should be grown in soils with a pH of between five and seven.

Pigeon Peas prefer to grow in full sun, but can also tolerate partial shade. They do not tolerate extreme (>40°C) temperatures very well.

Uses:

Fruit and Seeds:

The pods and seeds are harvested and are very nutritious. The pods can be eaten when green, and the seeds can be used when pods are green or brown. They may be eaten raw or cooked. Sprouted seeds are delicious.

Leaves:

The leaves can be eaten in small amounts as an herb, but are usually fed to animals.

Wood:

The dried stalks are commonly used as a source of fuel, but may also be used as thatch for basketry and other crafts.

Miscellaneous uses:

Pigeon Peas are nitrogen-fixing and nodulate with many types of *Rhizobia*. Generally, Pigeon Pea plants do not need to be inoculated with a special strain of bacteria to successfully fixate nitrogen. Nonetheless, Cowpea inoculants seem to be effective. Due to their nitrogen-fixing abilities Pigeon Peas are often interplanted with other crops as a support species, from which green manure can be pruned and allocated to other plants. Pigeon Peas can also be shaped into hedges.

Propagation:

Seeds usually germinate within two weeks when planted into a friable soil at a depth of 2-5cm. No pretreatment is necessary, but seeds can be soaked in water for 12-24 hours before planting.

Potential Drawbacks:

Pigeon Peas can exhibit a wide-range of pathogenic and pest problems. Most of these problems, however, are associated with more humid climates.

Nutrition:

Dehusked seeds (per 100g):	
Moisture	15.2%
Protein	22.3%
Fat	1.7%
Mineral Matter	3.6%
Carbohydrates	57.2%
Calcium	9.1%
Phosphorous	.26%
Carotene evaluated as Vitamin A	220 IU
Vitamin B1	150 IU

Fresh green forage:	
Moisture	70.4%
Crude protein	7.1%
Crude fiber	10.7%
N-free extract	7.9%
Fat	1.6%
Ash	2.3%

Seed oil:	
Linolenic acid	5.7%
Linoleic acid	51.4%
Oleic acid	6.3%
Saturated fatty acids	36.6%

Immature Seeds (per 100g)	
Calories	117 calories
Moisture	69.5%
Protein	7.2g
Fat	.6g
Carbohydrates	21.3g
Fiber	3.3g
Ash	1.4g
Calcium	29mg
Phosphorous	135mg
Iron	1.3mg
Sodium	5mg
Potassium	563mg
Beta-carotene equivalent	145µg
Thiamine	.4mg
Riboflavin	.25mg
Niacin	2.4mg
Ascorbic acid	26mg

Amino acid composition:	
Arginine	6.7%
Cystine	1.2%
Histidine	3.4%
Isoleucine	3.8%
Leucine	7.6%
Lysine	7%
Methionine	1.5%
Phenylalanine	8.7%
Threonine	3.4%
Tyrosine	2.2%
Valine	5%
Aspartic acid	9.8%
Glutamic acid	19.2%
Alanine	6.4%
Glycine	3.6%
Proline	4.4%
Serine	5%

Source: *Handbook of Energy Crops*

Related Species:
Atylosia cajanifolia is the closest wild relative. Another relative, *Cajanus kerstingii*, grows commonly in West Africa.

Caper

Scientific name: *Capparis spinosa*
Family: Capparaceae
Synonym: Caper Bush
Zone: 8b

Overview:
The Caper Bush is a small, spiny shrub growing up to one meter (3.2 feet) tall with a spread of two meters (6.5 feet). It is a winter-deciduous plant with an extensive root system, and is well-adapted to Mediterranean climates. The plant is best known for its flower buds, which are pickled for commerce. These flower buds bloom into white to pink flowers, which give rise to green-red fruits. The shrubs remain productive for at least 30 years, and may be annually pruned, because the flowers are born from the current season's growth.

Habitat and Requirements:
The Caper is well known and very common throughout the Mediterranean Basin. Nonetheless, it is thought to be native to the Middle East, because the plant can be found growing naturally as far east as northern India.

Since the Caper Bush is well-adapted to a Mediterranean climate it grows in areas with a cool, wet winter and a long dry season concentrated in the summer. The plant grows in areas where no rain may fall for nine or ten months out of the year. It is adapted to growing in areas that receive as little as 200mm rainfall per year, yet produces more reliably in areas that receive between 350-700mm of average annual rainfall.

The plant grows in very poor, but mostly well-drained soils. It seems well-adapted to rocky soils with very little organic matter. Nonetheless, it can tolerate clay loams as well. Due to its adaptation for growing in rocky areas, it is a common pioneer plant found growing between the rocks and stones of historical monuments throughout the Mediterranean region.

The caper is tolerant of summer heat of well over 40°C in the summer. In winter, however, the plant is only mildly hardy. The foliage starts to burn at temperatures in the upper twenties Fahrenheit. Therefore, foliage is only hardy down to zone 9b. The rest of the plant, however, is hardy down to zone 8b.

Figure 4: *C. spinosa* fruit. Sinai, Egypt.
Credit: Florian Prischl

Uses:

Fruit and Seeds:

Although less commonly implemented throughout the culinary world, the fruit of the Caper Bush, known as Caper berries, can be pickled and eaten in the same manner as the flower buds.

Flowers:

The flower buds are well-known commercially, and are usually pickled and eaten raw, especially in Italian cuisine. The flowers are also edible.

Leaves:

The leaves are even more seldom used than the fruits, but are also edible. They can be pickled or boiled, and are sometimes used as an ingredient in salads.

Miscellaneous uses:

Capparis spinosa has been found to grow in association with mycorrhizal fungi and nitrogen-fixing bacteria. These associations allow *Capparis spinosa* to pioneer very nutrient-poor soils. The plant's nitrogen-fixing abilities suggest that the Caper Bush might be a good green manure plant, whose prunings could be allocated towards other nearby plants in the form of rough mulch.

Propagation:

Propagation is best achieved with fresh seed. Fresh seeds should germinate within one month. Older seed is more difficult to germinate because it has usually dehydrated to some degree. Older seeds need to be cold stratified by being placed in a damp cloth in a refrigerator for two months. After two months, the seeds may be removed and placed in warm water and left to soak overnight before being planted in flats in the nursery.

Plants can also be propagated from stem cuttings, but are notoriously difficult to root. Furthermore, plants grown from cuttings tend to have weaker root systems.

Nutrition:

Pickled *Capparis spinosa* flower buds (per 100g):

	Nutrient Value:	Percentage of RDA:
Energy	23Kcal	1%
Carbohydrates	4.89g	4%
Protein	2.36g	4%
Total Fat	.86g	3%
Dietary Fiber	3.2g	8%

Folates	23mcg	6%
Niacin	.652mg	4.5%
Pantothenic acid	.027mg	.5%
Pyridoxine	.023mg	2%
Riboflavin	.139mg	11%
Thiamin	.018mg	1.5%
Vitamin A	138IU	4%
Vitamin C	4.3mg	7%
Vitamin E	.88mg	6%
Vitamin K	24.6mcg	20.5%

Source: USDA
National Nutrient Database

Sodium	2954mg	197%
Potassium	40mg	1%
Calcium	40mg	4%
Copper	.374mg	42%
Iron	1.67mg	21%
Magnesium	33mg	8%
Manganese	.078mg	3%
Phosphorus	10mg	1%
Selenium	1.2mcg	2%
Zinc	.32mg	3%
Carotene-ß	83 mcg	–

Related Species:

Related species include *Capparis sicula*, *C. orientalis*, and *C. aegyptia*. Furthermore, *Capparis spinosa* has many different varieties and cultivars, including *C. spinosa var. inermis*, which is a spineless variety.

Carob

Scientific name: *Ceratonia siliqua*
Family: Fabaceae
Synonyms: Locust Bean, St. John's Bread
Zone: 9a

Overview:

Carob is a small, evergreen tree growing up to a height of 10m (33ft) with an equal spread, and has thick, leathery leaves. Trees are dioecious, producing male and female flowers on separate trees, and are very drought-tolerant. Occasionally trees may be hermaphrodite, as is the case with the cultivar, 'Santa Fe.' The tree can live for hundreds of years, and remains in optimal production for about 90 years. Old trees develop very gnarled, thick, ornamental trunks. Male flowers produce a semen-like odor. The tree produces sweet, nutritious seed pods.

Habitat and Requirements:

Carob is native to the Mediterranean Basin around North Africa and the Middle East. As such, it is well-adapted to regions that have cool, wet winters followed by a long dry season with a hot summer. Trees can tolerate regular summer temperatures above 40°C. They have a large, extensive root system with a tendency to search for water in deep subsoils during periods of exceptional drought.

Although the tree is adapted to growing in areas where average annual rainfall is between 250-600mm, they may not set fruit in areas that receive less than 350mm, and may not produce a decent crop in areas that receive less than 500mm of rainfall. The application of a heavy mulch layer in late spring, however, can greatly improve fruit set in areas with lower rainfall.

The trees prefer well-drained, sandy soil, but are adapted to a wide range of soils, and are moderately salt-tolerant. The tree can also tolerate calcareous, alkaline soils, especially where limestone is the responsible agent for the high pH. The trees require full sun.

Figure 5: *C. siliqua*. Santa Barbara, California.
Credit: Mark Viergutz

Uses:

Fruit and Seeds:
The pods contain a high sugar content as well as a wide array of vitamins and minerals. The pods can be eaten raw, fermented to produce alcohol, made into a syrup or flour, and are often fed to animals as fodder. The pulp provides a sugar, which certain fungi cultures use to produce a protein-rich product. The flour can be used as a healthy chocolate substitute. A light-brown dye can also be extracted from the pods.

The seeds are edible and contain a high protein content and can be eaten by humans and livestock as well as by pets. The seeds are well known for containing a commonly used ingredient in the food industry, carob bean gum (CBG), also known as locust bean gum (LBG), which is useful as a thickening agent.

Wood:
The wood is close-grained and hard, making it an ideal material for utensils. It can also be used to produce high-quality charcoal, or firewood.

Miscellaneous uses:
Despite its categorization in the pea family, Carob has long been thought to be devoid of nitrogen-fixing capabilities. Recently, however, an unkown species of bacteria has been isolated from the root nodules of Carob. This bacteria is believed to be from the nitrogen-fixing genus, *Rhizobium*, and only grows when pH is above 7. These bacterial associations could further validate Carob's use as an agent for reclaiming degraded land.

Propagation:

Carob can be propagated by seed or from 2-3 year old cuttings, and are best grafted using shield budding. Budding is done when the root stock is one year old. Cuttings are difficult to root.

The boiling method, mechanical scarification, sulphuric acid and gibberellic acid are effective seed pretreatments. Seeds lose their viability after one year, and germination rates are considerably lowered after five years. Seedlings are very deep rooted, and some propagators even use a 3 foot section of pipe in order to accommodate the root. Other purists prefer to sow the seeds directly in the field in order to forego having to deal with the hassle in the nursery. Nonetheless,

Figure 6: Carob flour and pods.

many have propagated Carob in traditional plant containers without experiencing negative effects. Larger trees, however, are nearly futile to transplant, due to root complications. To be safe, Carob plants should not be left in the nursery for more than a year or two.

Potential Drawbacks:

Establishment of young trees could be a hassle, because roots tend to be deep, and gophers have a tendency to eat them. Rats may also girdle young trees. The ripe pods must be harvested before the onset of the rainy season, otherwise they will rot.

Nutrition:

Carob flour (per 100g):

Proximates:	
Water	3.58g
Energy	222kcal
Protein	4.62g
Total lipid (fat)	.65g
Carbohydrate, by difference	88.88g
Fiber, total dietary	39.80g
Sugars, total	49.08g

Minerals:	
Calcium	348mg
Iron	2.94mg
Magnesium	54mg
Potassium	827mg
Sodium	35mg
Zinc	.92mg

Vitamins:	
Vitamin C	.2mg
Thiamin	.053mg
Riboflavin	.461mg
Niacin	1.897mg
Vitamin B-6	.366mg
Folate, DFE	29mcg_DFE
Vitamin A, RAE	1mcg_RAE
Vitamin A, IU	14IU
Vitamin E (alpha-tocopherol)	.63mg

Lipids:	
Fatty acids, total saturated	.090g
Fatty acids, total monounsaturated	.197g
Fatty acids, total polyunsaturated	.216g

Source: USDA National Nutrient Database

Yeheb Nut

Scientific name: *Cordeauxia edulis*
Family: Fabaceae
Synonyms: Yeheb, Ye-eb, Yeeb, Yicib Nut Bush
Zone: 10b

Overview:

The Yeheb Nut is a small, evergreen shrub growing to a height of 2.5m (8ft) with an equal, horizontal spread. The shrub produces a taproot, which grows to a depth of at least 3m (10ft), and develops lateral roots closer to the surface of the soil. The shrubs can live for longer than 200 years, and can survive extreme droughts once they are established. Yeheb shrubs produce yellow flowers followed by small, 1-2 seeded pods that contain edible seeds known as Yeheb Nuts. The plant starts bearing fruit after it is about 3-4 years old.

Habitat and Requirements:

Yeheb is native to the area along the southeast border between Ethiopia and Somalia towards the Indian Ocean. This is a region that experiences an annual average rainfall of between 85-400mm. The region has two rainy seasons. One rainy season occurs between the months of March and June, and the other season occurs during October-November. Nonetheless, these rainy seasons are often unreliable and bring insubstantial, erratic rains followed quickly by dry conditions.

The Yeheb Bush has adapted to this climate and can even survive for two years without receiving any rain. Plants do not produce during drought years, but they sometimes go deciduous, making them able to survive harsh conditions until rains return. The plants grow in full sun in deep, red sands with very quick drainage. The soils where the plant grows have been measured to have a pH of 6.7–8.4. The plants do not grow near water courses, and are intolerant of waterlogging and frost.

Figure 7: *C. edulis.*
Credit: Asha Yahya

Uses:

Fruit and Seeds:

The shells of the seeds are easy to crack, and the seeds themselves are sweetish in flavor. The seeds are eaten raw, boiled, dried, and roasted, and taste like a cross between a Macadamia nut and a Cashew. The seeds are about the size of a Macadamia nut. The seeds are lower in protein and carbohydrates compared to most legumes, but are higher in fat and sugar, properties which contribute to their good taste. Roasted seeds can be used as a coffee substitute. The oil extracted from the seeds is edible, but does not taste nearly as good as olive oil. Yeheb Nut oil can also be used for soap-making.

Leaves:

The leaves are browsed by livestock and other wildlife. Additionally, the leaves can be used as a tea substitute. The leaves can be used to make a magenta-colored dye, due to the presence of cordeauxiaquinone.

Wood:

The wood is burned as fuel.

Miscellaneous uses:

The Yeheb Bush is nitrogen-fixing and can be used as a source of green mulch. The plant colonizes very nutrient-poor, sandy soils, and could be used as a pioneer plant in degraded areas.

Propagation:

Plants can be propagated from cuttings, but can be difficult to root. Seeds are only viable for up to one year, and generally lose their viability much sooner (about 4 months after harvesting). Since the seedlings concentrate on developing a deep taproot during the early stages of growth, seeds should not be grown in the nursery. The taproot is very fragile, and plants that have their taproot broken will die. Therefore, seeds should be direct sown in the field at a depth equal to that of the seed's diameter. Most fresh seeds will germinate within two weeks. Plants should be given supplemental irrigation during their first two years of growth in order for them to become established.

Potential Drawbacks:

The entire plant contains cordeauxiaquinone. The leaves have the highest concentration, and goats that eat the plant have been known to have their bones and teeth stained a reddish color. Other than this effect, no known negative effects have been documented with the consumption of cordeauxiaquinone.

Nutrition:

(Per 100g):

Seed:	
Water	7.8 - 16.9%
Fat	9.9 - 13.4%
Crude Protein	10.8 - 15.9%
Starch	31.4 – 41.5%
Sugar	12.4 – 24.9%
Reducing Sugar	2.5 – 5.3%
Sucrose	11.6 – 21.6%
Ash	2.2 – 3.8%
Calorific Value	446 Kcal/kg

Other amino acids:	
Aspartic acid	18.7
Serine	6.7
Glutamic acid	20.8
Proline	8.8
Glycine	8.2
Alanine	8.6

Essential Amino Acids:	
Lysine	3.9
Histidine	1.3
Arginine	9.1
Threonine	2.1
Valine	1.3
Methionine	2.4
Isoleucine	Trace
Leucine	3.8
Phenylanine	1.4
Tryptophan	0

Complementary amino acids:	
Cystine	Trace
Tyrosine	3.0

Source: *Non-timber Uses of Selected Arid Zone Trees and Shrubs in Africa*

Related Species:

Stuhlmannia moavi is a similar species from Tanzania.

Honeybush

Scientific name: *Cyclopia genistoides*
Family: Fabaceae
Synonym: Honeybush Tea
Zone: 8b

Overview:

Honeybush is a small, evergreen shrub growing to 1m (3.2ft) tall. It has dark, needle-like leaves that grow in clusters of three along the branches, similar in appearance to certain species of pine. Honeybush is not a very striking plant except in the spring when it reveals its profuse blooms of yellow flowers. These flowers give rise to brown pods, which contain the small seeds. The pods split open when the seeds are ripe. The plant is similar in appearance to Rooibos, and is utilized in much the same way. After a few years, plants will become woody, and should be cut back once a year when harvesting. The plant is called Honeybush because the flowers smell sweet, and the tea has a sweet taste.

Habitat and Requirements:

Honeybush is adapted to the Fynbos region of the West Cape of South Africa where it grows in an area that receives an annual rainfall of between 300-600mm. It grows in coastal and mountainous areas. Its habitat is considered a Mediterranean climate that has cool wet winters, followed by a long dry season lasting the majority of the year, and concentrated during the summer months.

The plant grows in full sun and sandy soil. It grows best in areas of considerable soil acidity where pH measurements range from 3.5-5. The plant germinates well after fire.

Uses:

Leaves:

The leaves are prepared and used in much the same way as the leaves of Rooibos. They are stripped from the branches, moistened and left in a heap to ferment. Then they are placed in the sun to dry. The leaves prepared in this manner are steeped as a tea that tastes sweeter than Rooibos tea.

Miscellaneous uses:

Since the plant is nitrogen-fixing, it can be used in land reclamation projects, or as a support species for other plants growing nearby. Since the plants are often burned, and resprout from their base following fire, the plants can also be coppiced.

Propagation:

Plants can be propagated from stem cuttings as well as from seed. Seeds need to be scarified. This can be done using sulfuric acid or simply with sandpaper. Since the Honeybush has evolved to live in a fire-prone habitat, the seeds also benefit from a smoke treatment. The seeds do not need to be exposed to fire, smoke alone is sufficient to spur germination.

Seeds should be sown just below the surface of a well-drained, highly acidic potting mix in late summer or fall. The plants should be grown in the nursery for one year, and then planted out at the start of the following rainy season.

Potential Drawbacks:

Honeybush cultivation outside of South Africa has either not been well-documented or executed successfully. Honeybush plants require good drainage and a very low soil pH. In order to cultivate and propagate Honeybush plants successfully, in many areas copious amounts of sand and pine bark fines should be combined to increase percolation rates while driving down pH levels to 4-5.

Aside from its low pH requirements, the lack of successful Honeybush cultivation outside of South Africa might be due to soil-specific nitrogen-fixing bacteria and/or mycorrhizal fungi absent in non-native soils. More investigations need to be carried out involving Honeybush's relationship(s) with symbiotic soil bacteria and/or mycorrhizal fungi.

Nutrition:

Honeybush has well-researched health benefits similar to those of Rooibos. Honeybush has a well-balanced mineral content that includes: Iron, Potassium, Calcium, Copper, Zinc, Magnesium, Manganese and

Figure 8: *C. genistoides*. Kogelberg Mountains, South Africa.
Credit: Ronnie Hazell

Sodium. Honeybush has also been shown to prevent cancer and lower cholesterol.

Related Species:

There are many species of *Cyclopia* that are used as Honeybush tea. These species include: *Cyclopia galioides*, *C. intermedia*, *C. subternata*, *C. sessiliflora*, and *C. maculata*.

Goumi

Scientific name: *Elaeagnus multiflora*
Family: Elaeagnaceae
Synonyms: Gumi, Cherry Silver Berry
Zone: 4a

Overview:

Goumi is a deciduous shrub growing up to three meters (10 feet) tall with a spread of two meters (6.5 feet). The plants flower in the spring and set fruit in midsummer. Flowers are hermaphrodite, so that only one plant is necessary for fruit set. The fruits are deliciously sweet with tart overtones and the occasional presence of astringency especially when under-ripe. Ripe fruits are about 1cm in diameter and are bright red.

Habitat and Requirements:

Goumi is native to China and Japan, and is well adapted to many dry, temperate climates around the world. Goumi can be grown in temperate regions of high rainfall with summer rain, as well as in drier Mediterranean climates without summer rain. It is wind-resistant, drought-tolerant, tolerates sea spray, and grows well in areas that receive 400-1200+mm of annual rainfall.

Goumi plants are hardy to zone 4a, but they will be killed back to their roots in this zone. Plants are fully hardy to zone 4b. The plants can be grown in full sun as well as in partial shade such as in the understory of forests with light canopies. The plants prefer well-draining

Figure 9: *E. multiflora*. Southern France.
Credit: André Abrahami

soil that is not very fertile.

Uses:

Fruit and Seeds:
The fruits and seeds are edible. The fruits can be eaten raw or cooked and are especially popular in pies or jams.
Leaves:
The leaves and stems can be used as fodder for livestock.
Miscellaneous uses:
The Elaeagnaceae family is a family of nitrogen-fixing trees and shrubs. Goumi is no exception, and, like the other members of its family, forms a symbiotic relationship with the genus of actinomycete bacteria known as *Frankia*. This association helps Goumi colonize degraded soils. Also, the plants can be coppiced and are commonly inter-planted in orchards to help provide extra nitrogen to the nearby orchard trees.

Propagation:

Goumi plants can be grafted onto each other, or onto other *Elaeagnus* species to use them as rootstock. Goumi can also be propagated from heel cuttings using the wood of the current year's growth. Cuttings should be taken after the plant has gone deciduous in the fall. Cuttings may take up to 12 months to properly root.

Propagating Goumi from seed also requires patience as seeds may take up to 18 months to germinate. Older seeds can take even longer to germinate. Nonetheless, germination can be hastened by following a stratification regimen of one month of warm stratification followed by three months of cold stratification. Seeds should be sown in flats and later pricked out and potted up or planted out in the field.

Potential Drawbacks:

Goumi has been shown to be invasive in many humid, temperate areas. In drier regions, however, there is less of a chance for Goumi to be invasive.

Nutrition:

Goumi fruit has been proven to have anti-cancer effects. The fruits are reported to contain essential fatty acids as well as flavanoids, other bioactive compounds, and vitamins A, C, and E.

Related Species:

Elaeagnus angustifolia, known as the Russian Olive, is a temperate tree species with edible fruit. *E. philippinensis* is a shrub with edible, pink fruit that grows well in tropical and subtropical areas, and is hardy to zone 9b.

African Winterthorn

Scientific name: *Faidherbia albida*
Family: Fabaceae
Synonyms: *Acacia albida*; Apple-Ring Acacia, Camel Thorn, White Thorn
Zone: 9b

Overview:

The African Winterthorn is closely related to African Acacia species, and indeed was once considered to belong to that genus. The African Winterthorn is a fast-growing (up to 1.5 meters/year) deciduous tree that generally grows to 15m in height with a 10m (33ft) spread. In moister areas the trees can grow up to 30m (98ft) in height. The plant is bipinnately compound with small, bluish green leaflets that help create a humus-rich litter layer beneath their canopy. It has small thorns growing in pairs along its branches.

The African Winterthorn is a very unique deciduous tree, due to the inverted phenology it exhibits in comparison with similar arid and semi-arid species. The plant is deciduous during the rainy seasons instead of the dry seasons. African Winterthorn trees lose their leaves at the onset of the rainy season and leaf out, bloom and fruit during the time of the year when there is no precipitation.

Habitat and Requirements:

African Winterthorn has a very wide distribution range, occurring throughout most of Africa and into the Middle East. This indicates that African Winterthorn is adaptable to many different climates. The

Figure 10: *F. albida* seed pods. Yuma, Arizona

tree loses its leaves at the start of the rainy season, and loses its leaves twice in areas where there are two rainy seasons. In Israel the tree loses its leaves during the winter rainy season, and in areas of East Africa and southern Africa the tree loses its leaves during the summer rainy season.

African Winterthorn grows in climates with mean annual rainfalls of between 200-1500mm. The tree seems to perform best in climates with at least 650mm of average annual rainfall, but is also extremely deep rooted and can grow at the lower end of this rainfall range in areas where there is groundwater. A mature tree's taproot can grow to at least 15m (49ft) deep, and, in some areas, nitrogen-fixing root bacteria have been located at the water table where the water table is 35m (115ft) deep. Root depth is generally determined by the availability of moisture.

The tree tolerates a wide variety of soil types, but prefers sandy alluvial soils, and generally avoids clay soil. African Winterthorn grows along seasonal and perennial watercourses and also on the interfluves of rivers in slightly higher areas that are not prone to flooding. Individual specimens have been known to survive temperatures down to -5°C (23°F). The trees grow in full sun.

Uses:

Fruit and Seeds:

The seeds are boiled and provide a protein-rich famine food. The trees start setting seed when they are about eight years old. When mature, five trees will produce about 83kg (183lbs.) of crude protein per year from their pods. Although the seeds can be cooked and eaten by humans, the pods are usually served whole as fodder for livestock.

Leaves:

The leaves are used as protein-rich fodder for livestock.

Wood:

The wood is used to make furniture and tools.

Bark:

The bark can be chewed on as a toothbrush substitute, and is said to prevent cavities due to the presence of fluorine.

Miscellaneous uses:

These trees can pioneer degraded soils by their associations with *Bradyrhizobium* and mycorrhizal fungi. The plants allow light to penetrate their canopy during the rainy season when they have no leaves and quickly create humus with the breakdown of their fallen leaves by the rain. During the dry season, plants create a light canopy and foster a shady, humid microclimate by extracting moisture from the water table. Due to these factors, crops such as sorghum, corn and millet are often interplanted amongst African Winterthorn woodlands.

The trees can also be coppiced and pollarded after they lose their leaves. The branches can be lopped off and used as fencing.

Propagation:

Seeds can be planted, and left for up to six months, in air-pruning containers that are at least 30cm deep. Nonetheless, seeds should be direct sown at a depth of 15mm, since the taproot develops quickly and could easily lead to complications in the nursery or when transplanting. When the seedling is about 8cm tall it has a taproot that is about 70cm deep.

If kept cool and dry, seeds have a long shelf-life, and don't seem to lose any viability after three years. The best pretreatment for seeds is to nick them with a nail clipper. With larger quantities of seed, however, it may be more practical to soak them in sulphuric acid or submerge them in boiling water, and then leave them in the water to cool. Most seeds will germinate within one month.

Potential Drawbacks:

African Winterthorn is prone to rot in areas of waterlogging.

Nutrition:

Whole pods:	
Moisture	7.7%
Crude Protein	12.2%
True Protein	9.8%
Fat	0.9%
Carbohydrates	54.9%
Crude Fiber	25.7%

Seeds:	
Moisture	9.4%
Crude Protein	27.2%
True Protein	19.7%
Fat	1.9%
Carbohydrates	45.3%
Crude Fiber	12.2%

Source: *Non-timber Uses of Selected Arid Zone Trees and Shrubs in Africa*

Honey Locust

Scientific name: *Gleditsia triacanthos*
Family: Fabaceae
Synonyms: Common Honey Locust, Sweet Locust, Honey Shuck
Zone: 4a

Overview:

Honey Locust is a large, deciduous tree often covered with large spines. On average, mature trees grow to about 20m (66ft) in height with a 15m (49ft) spread. This tree is widely planted throughout temperate regions of the world for its ornamental value as well as for economically valuable features. While trees can be dioecious (male and female flowers on separate trees), they also commonly have perfect flowers and/or separate male and female flowers on the same tree. Female flowers give rise to large edible pods with sweet pulp and edible seeds. Trees start producing when they are around 10 years old, and remain in optimal production when they are 25-75 years old. Honey Locust trees live to be about 125 years old.

Habitat and Requirements:

Honey Locust is native to the USA in the states near the Mississippi river. This is a naturally humid region with summer rains. Nonetheless, Honey Locust is adaptable to drier regions and is easily grown throughout the Great Plains region to the west. Honey Locust naturally grows in areas that receive between 510-1250mm of average annual rainfall. Honey Locust can grow in drier areas, but needs supplemental irrigation or access to groundwater. In coastal areas with cooler summers pod yields are reduced.

Honey Locust prefers moist sites with alluvial soils and does not perform as well in clay or gravelly soils, or areas with shallow soil. Trees prefer to grow in areas where the soil pH is between 6 and 8, but trees are also found growing in acidic soil where pH may be as low as 4. Trees tolerate salinity, and have an adaptable root system, which makes them drought-tolerant. Trees are very shade-intolerant, and typically establish themselves in open areas where there is no tree canopy.

Uses:

Fruit and Seeds:

The pods and seeds are edible. The seeds can be roasted and ground as a coffee substitute. The pods contain a sweet pulp that can be eaten or used to make a drink. Pods may be ground up and eaten whole. The pods and fruit are also commonly fed to livestock. Trees are known to be very productive, and a single mature specimen can produce about 40kg of pods annually.

Flowers:

Flowers can be eaten, and yield an excellent tasting honey in apiculture.

Leaves:

The leaves should not be eaten by humans, but are an excellent, high-protein, livestock fodder.

Figure 11: The canopy of a mature *G. triacanthos*. Santa Barbara, California

Wood:
Honey Locust yields a high-quality fuelwood. The wood, however, is better used in construction and craftsmanship, and has even been used for guitar-making.
Miscellaneous uses:
It is debated whether or not Honey Locust fixates nitrogen. Most sources suggest that Honey Locust does not fixate atmospheric nitrogen due to the absence of any root nodulation. However, one cluster of studies conducted by Bryan et al. has yielded evidence that Honey Locust does indeed fixate nitrogen. In any case, Honey Locust is usually able to grow well in poor soils. The tree coppices and pollards well, and can be kept as a small barrier hedge.

Propagation:

Honey Locust can be grafted in order to select ideal cultivars. This is best done by bud grafting young trees during summer. In order to propagate trees from seed, seeds should be scarified by mechanical methods, sulfuric acid soaking, or by soaking in very hot water. Additionally, the seeds can be soaked for one or two days prior to planting. Using fresh seeds or properly stored seeds, the seedlings should germinate within one month with a germination rate of at least 75%.

Potential Drawbacks:

The seeds are often attacked by weevils. Pods and seeds should be stored at freezing temperatures to preserve their freshness and to prevent seed weevils from thriving.

Nutrition:

Seeds (per 100g):	
Protein	10.6 – 24.1%
Fat	0.8 – 4.3%
Carbohydrates	84.7%
Fiber	21.1%
Ash	4%
Calcium	280mg
Phosphorous	320mg

Source: *Medicinal Plants of China*

Related Species:

Gleditsia triacanthos var. inermis is a common spineless variety.

False Mopane

Scientific name: *Guibourtia coleosperma*
Family: Fabaceae
Synonyms: Large False Mopane, African Rosewood, Rhodesian Mahogany, Copalwood
Zone: 10a

Figure 12: *G. coleosperma*. Hwange National Park, Zimbabwe. Credit: Bart Wursten

Overview:

False Mopane is a large, nearly evergreen tree growing up to 20m (66ft) tall with an equal spread. The trunk is made up of very dense wood, and has pinkish bark that often flakes off in large pieces leaving dark patches along the trunk. The star-shaped flowers are white and give rise to small, woody, capsule-like pods about 2-3cm in diameter. Upon ripening, the pods split to reveal a brown seed that is covered in a bright red aril. Both the seed and the aril can be eaten.

Habitat and Requirements:

False Mopane is native to the Kalahari Basin at the northern end of the Kalahari Desert. The tree is found growing in Angola, Zambia, Namibia, Botswana, Zimbabwe, and even as far north as the Democratic Republic of Congo. Throughout the majority of its habitat, False Mopane experiences an average annual rainfall range of between 300-700mm with up to 90% of this rain occurring during the summer months from November – March.

The tree requires very well-drained soil, and is almost always found growing on deep Kalahari sands. The roots grow very deep in an attempt to locate groundwater.

Uses:

Fruit and Seeds:

When the pods split, they reveal a seed covered in a very showy, bright red aril. The arils are oily and can be consumed along with the seed. It is a very important famine species throughout its native habitat since the seeds and arils are eaten raw or roasted and probably provide healthy levels of protein and fat.

Wood:

The wood is very dense and resists abrasion, making it useful for wood flooring. It is also esteemed for furniture-making, general construction, paper-making, and for carvings.

Miscellaneous uses:

False Mopane trees are nitrogen-fixing and can be used on relatively infertile, sandy soils as a pioneer tree to help reclaim the land.

Propagation:

The seeds require no special pretreatment, but can be soaked in water overnight before planting. Germination rates are good, and most seeds germinate within two months from planting. The seeds should be direct sown in the field since they are very deep rooted. They do not take well to container culture, and even with a well-draining potting mix, the seedlings hardly surpass their cotyledon stage. Plants grow very poorly in sites that are not predominantly sand.

Potential Drawbacks:

The tree is not practical to raise in a nursery setting. Also, the trees suffer in heavy soil. Avoid planting in areas where there is a high presence of clay in the soil profile.

Leucaena

Scientific name: *Leucaena leucocephala*
Family: Fabaceae
Synonyms: *Leucaena glauca*; Miracle Tree, Guaje, Leadtree
Zone: 9b

Overview:

Leucaena is a self-fertile, evergreen tree that is widely utilized in agroforestry systems. Leucaena trees are fast-growing, reaching a maximum height of 20m (66ft) in as little as six years. The trees are deep-rooted with taproots growing to at least 5m (16ft) deep. Under ideal conditions, some trees may produce seeds within as little as one year. Generally however, the trees are slower growing and less productive in dry climates than in humid climates.

Habitat and Requirements:

Leucaena is thought to be native to the Yucatan Peninsula in Mexico and south into parts of Central America. Therefore, the tree is well adapted to warm, humid weather with ample rainfall. Surprisingly, however, the plant is also very drought-resistant and can grow in much drier conditions, thanks to its deep roots that tap ground water. The trees naturalize in areas that receive between 500-2000mm of average annual precipitation.

The plant prefers to grow in full sun, and tolerates a wide range of soil conditions, but prefers to grow in deep clay loam soil, and does not tolerate waterlogging. The trees are moderately salt-resistant,

Figure 13: *L. leucocephala*. Diamond Head Crater, Hawaii. Credit: T.M. Stumpe

aluminum-resistant, wind-resistant, slightly frost-tolerant, can be grown on steep slopes, and have a pH tolerance of 4.3-8.7.

Uses:

Fruit and Seeds:

The unripe pods and seeds can be cooked and eaten by humans. The tree produces copious amounts of seeds (up to 2 tons/hectare under ideal conditions).When raw they contain a toxic compound, mimosine, which is mostly destroyed when cooked. Nonetheless, Leucaena should only form a minor part of one's diet. The pods and seeds can also be eaten by livestock, however, non-ruminant animals are affected by the toxicity in the same way as humans, so consumption should be limited.

Leaves:

The foliage is similar in nutrition content to alfalfa and is popular as cattle fodder.

Wood:

The wood is a good source of general lumber and is often used for posts and furniture. Also, it is a choice fuelwood because it burns slowly without emitting much smoke, and can be made into charcoal.

Miscellaneous uses:

Due primarily to its nitrogen-fixing properties, Leucaena is a very useful tree for any productive system. It is often used to reforest degraded land, and can be planted as an overstory for shade-tolerant crops. Leucaena trees can also be coppiced and the vegetation used as mulch for other plants nearby. The trees may be pruned into hedges and used as support for vines.

Propagation:

Propagation is best achieved by seed. Fresh seeds should be used, since seeds appear to lose viability over the course of months or years. To achieve the best germination rates, seeds should be scarified. Scarification can be accomplished by mechanically nicking the seed coat, soaking in a sulfuric acid solution, or soaking for a couple minutes in boiling water. The seeds can then be sown into a seed flat. Most seeds germinate within a month and can be picked out when they develop their true leaves. They then prefer to be grown in narrow pots about 15cm deep with bottom air-pruning for one season before being planted out in the field.

Potential Drawbacks:

Mimosine is a non-protein amino acid that is present throughout Leucaena, has mild insecticidal properties, and is toxic to humans and non-ruminant animals. Consumption of high levels of mimosine can lead to hair loss, and an enlarged thyroid as well as other health complications.

Nutrition:

Raw pods (per 100g):	
Calories	59
Moisture	80.7%
Protein	8.4g
Fat	.9g
Total Carbohydrates	8.8g
Fiber	3.8g
Ash	1.2g
Calcium	137mg
Phosphorous	11mg
Iron	9.2mg
B-carotene equivalent	4,730mg
Riboflavin	.09mg
Niacin	5.4mg
Ascorbic Acid	8mg

Source: *Handbook of Energy Crops*

Related Species:

Leucaena pulverulenta reportedly contains less mimosine.

Blue Lupin

Scientific name: *Lupinus angustifolius*
Family: Fabaceae
Synonym: Narrow-Leaved Lupin
Zone: 9a (Plants can be grown in much lower zones as long as they have at least 5-6 months of temperatures above -6°C)

Overview:

Blue Lupins are small, herbaceous plants commonly cultivated as ornamentals for their beautiful, bright blue flower stalks. They can grow up to 1m (3.2ft) tall with a lateral spread of about a third of a meter (1ft). They are annuals, which germinate during the winter, flower in the spring and finish fruiting by the end of the summer. Aside from their ornamental value, the seeds of Blue Lupin provide a protein-rich source of food.

Figure 14: *L. angustifolius*. Ourense, Spain.
Credit: Asociación BIGA

Habitat and Requirements:

Blue Lupin is native to the Mediterranean basin and thrives in areas of full sun that receive from 350-1,660mm of average annual precipitation.

The plant prefers to grow in areas with sandy or sandy loam soils that are deficient in nitrogen, and are well-adapted to growing in these areas because of their deep taproots (up to 2m deep), and their ability to fix nitrogen. Blue Lupins tolerate a soil pH range from 4.9-8.2, but generally do best where soil pH is less than 7.

Uses:

Fruit and Seeds:

A distinction is made between bitter and sweet Lupins. Bitterness is an indicator of the presence of toxins in the seeds. Nonetheless, "sweet" Lupin seeds should not be consumed raw. In all cases, the seeds of Blue Lupins should be soaked in water overnight, and then rinsed out. Then the seeds are soaked in a new pot of water and that water is discarded and replaced once or twice during the cooking process. Sprouting the seeds also apparently helps diminish their toxic alkaloid content. Under ideal conditions, one hectare of Blue Lupin plants can yield 500-600kg of seed in a single year.

After cooking, the seeds can also be pounded into flour. This flour, when added to wheat flour at the rate of 20%, creates a complete protein and extends the shelf life of the flour.

Flowers:

Blue Lupin flowers attract bees and beneficial insects.

Miscellaneous uses:

Blue Lupin is nitrogen-fixing and is a useful crop for helping to reclaim depleted, sandy soils. Blue Lupins, like most other Lupins, require the presence of specific strains of bacteria in order to fixate nitrogen. Blue Lupin should be inoculated with strains of *Rhizobium lupini*. In areas where Blue Lupin has been present, continued inoculation with *Rhizobium* is unnecessary. Due to the root exudates from Blue Lupins, plants growing nearby benefit from the Phosphorous that is leached from the rocks and other solids in the soil.

Propagation:

Seeds can be soaked in water overnight before sowing. They should then be planted at a depth of 2.5-5cm and are best direct-sown out in the field, but may also be planted in flats and pots. Most seeds will germinate within two weeks. Seeds stored for up to two years do not appear to lose any of their viability. Germination rates are very good. Occasionally, however, seeds will be infected by fungi and should be discarded. Direct seeding should be done in late fall before the onset of winter storms.

Potential Drawbacks:

The seeds can be attacked and ruined by toxic fungi. This is more common in areas that receive rain during the period of seed ripening. Edible Lupin species do not withstand waterlogging and do not perform very well in heavy, clay soils.

As mentioned earlier, Blue Lupin seeds contain toxic alkaloids. Specifically, they contain d-lupanine, sparteine, lupanine, spathulatine, and hydroxylupanine. Seeds should be soaked in various changes of water to most effectively leach out the toxins.

Nutrition:

Blue Lupin seeds are protein-rich and nutritionally comparable to soybeans. Blue Lupin seeds also have cholesterol-lowering effects.

Amino acid content of Blue Lupin seeds (g/16 g N):	
Arginine	11.62
Histidine	2.57
Isoleucine	3.91
Leucine	6.90
Lysine	4.75
Methionine	0.66
Phenylalanine	3.85
Threonine	3.54
Tryptophan	1.00
Valine	3.78

Source: Department of Agriculture and Food: Government of Western Australia

Related Species:

Two other species are grown throughout the Mediterranean basin for food: *Lupinus albus* and *L. luteus*. Another species, *Lupinus mutabilis*, has an excellent nutritional profile and taste, and is grown throughout the cool Andean highlands of Peru and Bolivia.

Tepary Bean

Scientific name: *Phaseolus acutifolius*
Family: Fabaceae
Synonyms: Tepari, Yori-Muni, Pavi
Zone: 7a (The plant grows naturally as an annual during the warmer parts of the year, thus avoiding frost.)

Overview:

The Tepary Bean grows either as a small, herbaceous shrub up to about 76cm (30in) tall, or as a vine that spreads up to 3 or 4 meters (10-13ft). The plant grows as an annual primarily outside of the winter months. The plant takes from 70-85 days to reach maturity. It flowers primarily at two different times of the year: first in April-May, and then again from July-September. The flowers are purple, and yield 7-9cm long pods that contain about 5-6 beans per pod, which are best harvested when the pods are brown and not overly dry. Tepary Beans were cultivated by native people in the American Southwest along with Squashes and Corn as a part of the "Three Sisters" agriculture. In warm climates, the plants are usually grown twice in one year.

Habitat and Requirements:

Tepary Beans grow during the warmer parts of the year in order to avoid frost, and prefer rain or irrigation during their initial stages of growth. When the plant has matured beyond one or two months, however, the plant does not require any water in order to successfully flower and produce fruit. The plant will thrive in prolonged, extreme heat. In many cases, one monsoonal storm will suffice to grow Tepary Beans throughout their entire life cycle. The plant is native to the Sonoran Desert in North America, and is therefore well adapted to growing in arid and semi-arid regions that receive between 200-500mm of average annual rainfall, with the bulk of that rainfall occurring during the summer months.

Figure 15: *P. acutifolius*. Tucson, Arizona.
Credit: Jay Tracy

The plant can tolerate wetter conditions, but is prone to fungal diseases in areas with higher rainfall and high humidity. The plant requires a well-drained soil with a pH range from 5.5-8, and does not tolerate waterlogging. The plants grow well in moderately sandy areas to areas of clay loam. Tepary Beans dislike salty soil and salty irrigation water. They require full sun.

Uses:

Fruit and Seeds:
The seeds can be extracted from the ripe pods, and can be ground into flour or boiled and eaten. The plants produce anywhere from 1,100-4,000kg of seed per hectare. The beans have a sweet, nutty taste, and creamy texture when cooked.

Leaves:
The leaves, along with the rest of the plant, can be used as fodder for livestock, and are nutritionally comparable to alfalfa hay.

Miscellaneous uses:
Since the plants are nitrogen-fixing they can be planted to reclaim degraded soils, as well as planted to serve as nurse plants for other useful plants growing nearby.

Propagation:

Tepary Beans are very easy to propagate. The plants germinate readily. To ensure success the seeds should be soaked in water overnight, and then sown directly in the ground the following day. The plants associate with the *Rhizobium* strains that are used to inoculate Lima Beans (*Phaseolus lunatus*).

Potential Drawbacks:

The plants are intolerant of waterlogging and salty conditions.

Nutrition:

Tepary Beans are high in protein (usually about 23-30%) and, with regard to amino acid content, are almost a perfect complement to cereal grains. Tepary Beans also contain higher levels of oils and calcium, iron, magnesium, zinc, phosphorus, and potassium than most other beans. Also, compared to other beans they have lower levels of polyunsaturated fat and starchy compounds, which means they are easier to digest.

Mineral composition of beans (mg/100g):	
Calcium	182mg
Iron	12.1mg
Magnesium	204mg
Zinc	4.7mg
Phosphorus	480mg
Copper	1.3mg
Manganese	2.8mg
Potassium	1614mg
Sulfur	340mg

Source: Hamama, A.A. and H.L. Bhardwaj

Manila Tamarind

Scientific name: *Pithecellobium dulce*
Family: Fabaceae
Synonyms: Guamúchil, Madras Thorn, Opiuma
Zone 9b

Figure 16: *P. dulce* seed pod. Kamehameha, Maui, Hawaii.
Credit: Forest & Kim Starr

Overview:

This sweet, pod-producing member of the pea family is a popular food resource in Mexico, where it is widely planted. These evergreen trees have a quick growth rate, outpacing most Mesquite trees. Under ideal conditions, Manila Tamarind achieves a height of up to 10m (33ft) in 5-6 years. Eventually trees max out at a height of 18m (59ft) with a broad canopy adorned with spines, oblong leaves, white flowers, and eventually curled, green to bright red edible pods.

Habitat and Requirements:

Manila Tamarind is most likely native to central or southern Mexico, but is reported as being native from coastal California through Mexico and Central America, and throughout South America. The exact origin of this species is unknown, and was probably introduced to California and other regions by the Spaniards on their route from Mexico. Recently, it has naturalized in India, Africa, Hawaii, and many islands of the Caribbean. Manila Tamarind is widely adaptable, and grows in areas where annual rainfall is between 400-1650mm. Individual specimens have been known to survive temperatures down to -5°C (23°F). The trees tolerate nearly all soil types from clay to sand, and tolerate saline conditions. The trees grow best in full sun.

Uses:

Fruit and Seeds:
The pods and seeds can be eaten raw. They can be ground into a powder the same way mesquite pods are processed, or they may be blended into a drink or roasted. The pods and seeds are a choice animal fodder. While the seeds are edible, their oil may also be utilized in the soap-making process.

Flowers:
Bees love the flowers and the pollen produces a high-quality honey.

Leaves:
Although not a primary choice, the leaves may also be browsed safely by livestock.

Wood:
The wood can be used for construction and as firewood.

Bark:
The bark, which produces a yellow dye, may also be used for tanning. Wounds produce sap, which can be utilized in a similar way to gum arabic.

Miscellaneous uses:
Manila Tamarind trees coppice readily. Since coppicing tends to increase the occurrence of thorns, the trees can be planted densely and kept short in hedgerows to deter larger animals. Manila Tamarind is also a nitrogen-fixing tree, forming a symbiosis with *Rhizobium* bacteria, which helps trees improve the soil quality.

Propagation:

Seeds may be soaked overnight before being placed into trays and covered with a light potting mix at a depth of 2cm. Although nicking is usually unnecessary, it may also hasten germination. Most seeds germinate within two weeks. Seedlings have a small root system, and can be easily started in and transplanted from seed flats. They may also be directly sown into pots or, with less success, directly sown in the field. Fresh seed is preferred, as most seed loses its viability after a few years. Trees are also propagated easily from cuttings.

Potential Drawbacks:

This tree has proved to be rather weedy in parts of Hawaii, St. Croix and other Caribbean islands. Also, it is blown over easily in high winds in humid regions. This latter problem is due to the frequent availability of water, and is not a problem in dry climates where irrigation and rainfall is unavailable for months at a time.

Nutrition:

Pulp (per 100g):	
Water	75.8-77.8g
Protein	2.3-3g
Fat	0.4-0.5g
Crude Fiber	1.1-1.2g
Ash	0.6-0.7g
Carbohydrates	18.2-19.6g
Calcium	13mg
Phosphorous	42mg
Iron	0.5mg
Sodium	19mg
Potassium	20.2mg
Vitamin A	25 IU
Thiamine	0.24mg
Riboflavin	0.1mg
Niacin	0.6mg
Ascorbic Acid	133mg
Calories	79

Seed:	
Kernel	70.6%
Water	21.7-56.4%
Protein	10.5-29.9%
Fat	8.0-17.7%
Crude Fiber	3.9%
Ash	1.6-2.3%
Carbohydrates	19.6-28.4%

Seed Oil:	
Oleic Acid	51.1%
Linoleic Acid	24.0%
Saturated Acids	24.3%

Source: *Under-utilized Tropical Fruits of Thailand*

Argentine Mesquite

Scientific name: *Prosopis alba*
Family: Fabaceae
Synonyms: White Mesquite, White Algarrobo, Algarrobo Blanco, White Carob Tree, Ibope, Igope, Tacu
Zone: 9a

Overview:

Argentine Mesquite is a spiny, semi-deciduous species of Mesquite that can attain a maximum height of up to 15m (49ft) with a canopy spread of up to 10m (33ft) in diameter. The trees are self-incompatible, and require cross-pollination with a genetically distinct specimen in order to set fruit. Whereas many species of Mesquite are fully deciduous during the winter months, Argentine Mesquite will not lose its leaves except during heavy frosts. The leaves are bipinnately compound and feather-like. The yellowish flowers are produced abundantly in spring and give rise to many long, slender, sweet pods that contain 12-30 seeds each. The trees are very productive, and young trees at a height of only 4m have been measured to produce 1.75kg of pods per tree in a single year.

Figure 17: *P. alba*. Provincia del Chaco, Argentina.
Credit: María del Carmen Hermida

Habitat and Requirements:

Argentine Mesquite is native to the Gran Chaco ecoregion primarily in northern Argentina, and is adapted well to other similar climates around the world that receive approximately 100-500mm of average annual rainfall. The trees grow in lowland, riparian habitats, or generally wherever there is ample groundwater located 3-10m below the surface of the earth. The plants are deep-rooted, and are able to efficiently extract water from the deeper soil horizons. The plants tolerate occasional flooding, and are somewhat tolerant of waterlogging, but not for prolonged periods of time.

The trees tolerate nearly all soil types, but are particularly fond of clay soils that contain lots of sand. They prefer a pH range of 6-8.5, but have also been known to grow well in areas where pH is as high as 10.3. The trees are also very salt-tolerant, and can withstand seawater salinity. The trees grow well in extreme heat.

Uses:

Fruit and Seeds:

The pods are placed through a blender, grinder or hammermill, and then pushed through a filter or screen to remove the larger fibers from the pods. The pericarp is what is sought after, and tastes like a cross between caramel and chocolate with smoky overtones. It has an excellent flavor. The resultant flour can be used for baking, to ferment an alcoholic beverage known as aloja, to produce ethanol as a fuel, mixed with water to form gruel, blended into smoothies, or added to other foods. It is very nutritious, and, when mixed with other grains such as wheat or oats, forms a complete protein.

The whole pods can also be used as fodder for livestock.

Flowers:

The profuse blooms make Argentine Mesquite a favorite tree for apiculture.

Leaves:

Foliage extracts have been proven to have antibacterial properties, and can be used along with bark, branches and gum to treat gastritis, and have antidysenteric, antiseptic, and emollient properties.

Wood:

The wood is durable and can be used for flooring and furniture. More often, however, it is used as fuel, since it burns for a long time. The wood of Mesquite trees is one of the most sought after for the production of charcoal.

Miscellaneous uses:

The sap the tree exudes can be chewed like gum, and has a sweet, smoky taste. This "gum" can also be used as a substitute for gum arabic.

The trees can be coppiced and pollarded, and regenerate very well. The trees fix nitrogen and should be inoculated with a *Rhizobium* strain specific to Mesquite trees. Due to their nitrogen-fixing capabilities they can be grown to help improve the soil of degraded land. Due to the fact that the trees tolerate moisture very well, and the fact that their canopies provide light shade as well as ample leaf litter, the trees are useful as a nurse species for growing typical, tender, annual crops in their understory to protect them from the climatic extremes of hot deserts.

Figure 18: Root-shoot ratio of one-month old *P. alba* seedling

Propagation:

Argentine Mesquite can be propagated from root cuttings and stem cuttings, but is most commonly propagated from seed. To increase germination rates, the hard seeds can be nicked. Otherwise, simply soaking them in water overnight will help. The seeds should be sown into flats at a depth of 1cm. Most seeds will germinate within the first month, and germination rates of at least 70% should be expected. The seeds retain viability for a long time. Even after 20+ years germination rates are still good. The seedlings are then potted up into deep pots, taking care not to damage their taproots. Plants should be planted out in the field after 2-3 months in order to avoid becoming root bound.

Potential Drawbacks:

Trees grown in the nursery for more than a couple years often develop girdled roots, and may not fully develop when planted out in the field. Such trees often blow over during storms.

Trees have been known to send out suckers. Frequency and extent of suckers tends to be correlated with the amount of pruning a tree receives.

Nutrition:

The pods of Argentine Mesquite contain 7-11% protein and 25-28% sugar. The pericarp alone contains (per 100g):

Water	4g
Protein	10g
Sugar	40g
Fiber	19g

Source: *Handbook of Energy Crops*

The following table lists nutrient data for the pericarp of a similar species from North America, Velvet Mesquite (*Prosopis velutina*):

Nitrogen	1.13%
Crude Protein	6.88%
Fat	2.24%
Fiber	23.48%
Ash	5.52%
Total Sugars	31.6%
Calcium	.63%
Magnesium	.09%
Sodium	.04%
Potassium	1.16%
Copper	6.4ppm
Zinc	9.6ppm
Manganese	11.6ppm
Iron	48.8ppm

Source: Uchytil, Ronald J.

Related Species:

Mesquite trees are almost always adorned with spines, yet occasional trees are spineless. These trees are commonly selected as cultivars for the urban landscape. Generally, however, spineless varieties are not as productive. Taxonomy among Mesquite species is difficult, especially considering that Mesquite species commonly hybridize with each other. There are 44 species of Mesquite, 20 of which are native to the Gran Chaco. The following are some other noteworthy species of Mesquite:

Prosopis nigra – Native to the more humid part of the Gran Chaco in Argentina and Paraguay.

Prosopis pallida – Native to the coastal deserts of northern Peru. Pods are often grown for an export market, which has helped keep native forests from being cleared for firewood.

Prosopis chilensis – Has a very similar distribution range as *P. alba*, but extends further south.

Prosopis velutina – Native to the Sonoran Desert in North America. Excellent pods as well as edible flowers.

Karoo Boer Bean

Scientific name: *Schotia afra var. afra*
Family: Fabaceae
Synonyms: Boerboon, Karoo Boerboon, Karoohuilboerboon, Hottentot Bean
Zone: 9a

Figure 19: *S. afra var. afra* flowers. Santa Barbara, California

Overview:

The Karoo Boer Bean is a small, evergreen tree growing up to 5m (16ft) tall with an equal spread. The leaves are bipinnately compound, dark green when old, and bronze-red when young. The tree blooms abundantly during February and March (August and September in the Northern Hemisphere). The flowers are bright pink, and produce copious amounts of nectar that often overflows from the flowers and drips onto the ground below. For this reason, the tree attracts many birds. The flowers give rise to explosive seed pods, which ripen in May and June (November and December in the Northern Hemisphere).

Habitat and Requirements:

Karoo Boer Bean trees are native to a small region of South Africa, encompassing the Klein Karoo, as well as the coastal areas of the southern part of the Western Cape, and the drier parts of the Eastern Cape. The distribution range of the Karoo Boer Bean receives approximately 250-800mm of average annual rainfall with a dry season concentrated in the summer.

Karoo Boer Beans are often found growing where the Jacket Plum (*Pappea capensis*) grows. The Karoo Boer Beans are mostly found growing along riverbanks, but the tree also grows in upland locations, and tolerates rocky soil. The trees prefer well-drained soil, and will be markedly stunted if planted in heavy clay soil.

Uses:

Fruit and Seeds:

The seeds are edible when green or when mature, but should be cooked. The roasted seeds can be ground into a meal.

Leaves:

The foliage can be used as fodder for livestock.

Bark:

The bark can be used as tannin after being ground and soaked in water.

Miscellaneous uses:

Like other members of the Pea family, Karoo Boer Bean has the ability to fix atmospheric nitrogen. It may be a useful species for land reclamation.

Propagation:

The seeds should be soaked in water overnight before planting. They can be sown in seed flats, and kept moist in partial shade. Most seeds will germinate within one to four months. The seedlings should be left in the seed flats until they have developed their third set of true leaves. At that time, they should be carefully transplanted into a deep pot in order to accommodate their disproportionately large taproots. Despite their taproots, the trees tolerate nursery conditions fairly well, and can even be kept as bonsai specimens. Ideally, the seeds should be started in the spring, and raised in the nursery throughout the summer and fall, and then planted out in the field to coincide with the winter rains.

Related Species:

All four species of *Schotia* produce seeds that can be eaten. *Schotia afra* is divided into two different varieties. Besides *var. afra*, the other variety, *S. afra var. angustifolia* is native to inland areas of Namaqualand up into southern Namibia, and is better adapted to more drought-afflicted, hot desert climates. This variety also produces edible beans, and tolerates freezes down to -8°C (18°F).

Aside from *Schotia afra*, other species include:

Schotia brachypetala – Native from the northern part of the Eastern Cape through KwaZulu-Natal, northern South Africa, Mozambique, and into Zimbabwe.

Schotia capitata – A coastal species growing indigenously from KwaZulu-Natal to Mozambique.

Schotia latifolia – Grows from the southern part of the Western Cape through the Eastern Cape, to Mpumalanga, and the Limpopo Province.

Tamarind

Scientific name: *Tamarindus indica*
Family: Fabaceae
Synonyms: Madeira Mahogany, Indian Date, Tamarindo, Taman, Dakar
Zone: 10a

Overview:

The Tamarind is a large, evergreen tree growing to 25m (82ft) in height with a canopy spread of up to 17m (56ft). This self-fertile species produces pinnately compound leaves, and hermaphroditic, bisexual flowers which turn into long, bean-like fruits. The fruits of Tamarind are cylindrical pods up to 20cm long containing 1-10 seeds surrounded by a matrix of sweet pulp. When grown from seed, a Tamarind tree usually takes 10-12 years to start bearing fruit. The trees usually live to be at least 200 years old, and remain productive for nearly all of those years.

Habitat and Requirements:

Tamarind grows as an indigenous tree in the savannas of Sub-Sahelian Africa where average annual rainfall ranges from 250-1500mm. Tamarind often shares habitats with the Baobab, *Adansonia digitata*. Tamarind seedlings are not frost-tolerant, and mature specimens are often lethally damaged when temperatures drop below -2°C (28°F). In some cases, mature trees sustain only minor damage when temperatures drop to -3°C (26.6°F), but this is not usually the case. Conversely, the trees tolerate extreme heat.

The trees are able to tolerate drought and extreme heat due to their deep, extensive root system, which taps groundwater. The trees are very wind-tolerant, as well as moderately salt-tolerant, and are therefore planted in seaside locations that are subjected to hurricanes. Tamarind trees prefer well-drained soil, and do not tolerate waterlogging. Their preference is to grow in deep, sandy, alluvial loam soils, but the plants also seem to do well in rocky areas, and on other nutrient-poor substrates. The plants require full sun at all stages of their life.

Uses:

Fruit and Seeds:

On average an individual mature tree yields 150 to 200kg of pods per year (12-16 tons per hectare). Half of the weight of the pods is due to the pulp content, which has a high sugar content and a sweet-sour taste. Fruits take about 6-10 months to fully ripen after the flowers have been pollinated.

The seeds are edible, and can be eaten roasted, soaked and boiled, and can also be fed to livestock. Although the seeds are edible to both livestock and humans, they are usually discarded along with the pods' shells in favor of the pulp, which is usually eaten raw. When immature, the whole pods can be eaten, but have a sour taste, and should be cooked. They are more often used as a spice.

Flowers:

The flowers can be eaten by humans and livestock. They also provide pollen for bees, which results in a good quality honey.

Figure 20: *T. indica.* Citrus grove Sand Island, Midway Atoll.
Credit: Forest & Kim Starr

Leaves:
The leaves can be used as fodder for livestock. Young leaves have a sour taste, and can also be used as a spice.
Wood:
The heartwood of Tamarind is strong, heavy, termite-resistant, and is used to make furniture. The wood is also an excellent fuelwood, and produces a very high-quality charcoal. The tree coppices well, therefore it can be sustainably managed for the production of fuelwood or charcoal.
Miscellaneous uses:
Tamarind is a nitrogen-fixing tree, and is therefore a potential candidate for a green manure crop and for reclaiming degraded land.

Propagation:

Seeds require no special pre-treatment, but may be soaked in water for 24 hours before being sown. In most cases, the seeds will germinate within two weeks, but may take up to two months. For fresh seeds, germination rates of 70-80% should be expected, although nicking the seed coat usually increases this rate. Shelf-life, however, is rather short, and seeds start to lose their viability in as little as 6 months. Within the first two months of growth, the seedlings produce a taproot up to 30cm in length, so the seeds should either be sown in very deep pots, or direct sown.

The plants can also be easily propagated via softwood terminal cuttings. The trees can also be vegetatively propagated by air layering as well as ground layering, and can be grafted using whip grafts, cleft grafts, as well as shield and patch budding.

Potential Drawbacks:

The tree produces deep shade, and may also have allelopathic effects. Either way, plants do not seem to be able to grow very well in the understory of Tamarind.

Nutrition:

Ripe pulp	(per 100g)
Calories	115
Moisture	28.2-52g
Protein	3.10g
Fat	0.1g
Fiber	5.6g
Carbohydrates	67.4g
Invert Sugars	30-41g
Ash	2.9g
Calcium	35-170mg
Phosphorus	54-110mg
Iron	1.3-10.9mg
Sodium	24mg
Potassium	375mg
Vitamin A	15 IU
Thiamine	0.16mg
Riboflavin	0.07mg
Niacin	0.6-0.7mg
Ascorbic Acid	0.7-3.0mg
Tartaric Acid	8-23.8mg

Young leaves	(per 100g)
Moisture	70.5g
Protein	5.8g
Fat	2.1g
Fiber	1.9g
Carbohydrates	18.2g
Ash	1.5g
Calcium	101mg
Magnesium	71mg
Phosphorus	140mg
Iron	5.2mg
Copper	2.09mg
Chlorine	94mg
Sulfur	63mg
Vitamin A	250mcg
Thiamine	0.24mg
Riboflavin	0.17mg
Niacin	4.1mg
Ascorbic Acid	3.0mg
Oxalic Acid	196mg

Flowers	(per 100g)
Moisture	80g
Protein	0.45g
Fat	1.54g
Fiber	1.5g
Ash	0.72g
Calcium	35.5mg
Phosphorus	45.6mg
Iron	1.5mg
Vitamin A	0.31mg
Thiamine	0.072mg
Riboflavin	0.148mg
Niacin	1.14mg
Ascorbic Acid	13.8mg

Source: *Fruits of Warm Climates*

Marama Bean

Scientific name: *Tylosema esculentum*
Family: Fabaceae
Synonyms: *Bauhinia esculenta*; Morama Bean, Gemsbok Bean, Tamami Berry
Zone: 7b

Overview:

The Marama Bean is a non-climbing, perennial, vine-like groundcover that produces runners up to 6m (20ft) long. The plant produces double-lobed leaves and bright yellow flowers, which typically appear during the months of December and January (June and July in the Northern Hemisphere). The flowers produce woody pods that contain one to six large, edible seeds. The seeds weigh approximately two-three grams each, and usually ripen within the pods during the month of April (October in the Northern Hemisphere). The plant is also valued for its large, edible, red-brown tuber, which becomes very large. One specimen was found in Botswana that weighed 277kg. The specimens are easy to locate, because they often form rings of rocks, which are unearthed by the swelling tuber.

Habitat and Requirements:

The Marama Bean is native to northern South Africa, Botswana, Namibia, Angola, Zambia and Mozambique. It is commonly found in the dry regions of the Kalahari and Namib deserts that occasionally experience years that receive virtually no rain. The plant is also found growing in more humid savanna regions of southern Africa, so the plant should be considered as being able to grow well in areas that receive between 100-800mm of average annual rainfall. The rains come during the summer months as monsoonal thunderstorms.

The plant requires sandy soil in order to grow optimally. In heavy soils, the tuber fails to develop, and the plants are subject to rotting. The plant prefers neutral to acidic soils, and is nearly always found growing in areas with deep sands. The fact that the plant displays winter dormancy, and has the ability to resprout from its tuber, means that Marama Bean is adaptable to more temperate regions, since it can survive frosts. Marama Bean should be considered cold tolerant down to at least -12°C (10°F). The plants also tolerate extreme heat very well, as their native habitat commonly experiences temperatures of around 50°C (122°F) during the summer.

Uses:

Fruit and Seeds:

The seeds are very nutritious, and, when roasted, have a taste similar to smoky, roasted cashews. They are sometimes eaten raw, but this is not advised due to their trypsin inhibitor activity. The seeds should be roasted within their shells, and then peeled before consuming. The seeds are also boiled or ground into a powder. The beans have a good shelf-life, and, if kept cool and dry, they can last for years.

The oil extracted from the seeds is a high-quality polyunsaturated oil similar to almond oil. It can be pressed conventionally or extracted using solvent methods.

Roots:

The tubers are commonly harvested when they weigh about 1kg. They are then roasted, and eaten. They are surrounded by a rather leathery skin, which should not be consumed. The tubers have been known to grow up to 277kg of which about 90% is water. Therefore, the tubers form an important source of emergency water.

Miscellaneous uses:

Since the plant is a ground cover that hugs the soil surface, it helps prevent erosion, making it a useful plant for combating desertification. All parts of the plant can be used as fodder for livestock.

Propagation:

The seeds should *not* be soaked before sowing. The excess moisture often causes the seeds to rot. Instead, the most effective pretreatment is to mechanically scarify the seed coat. The seeds can be sown direct or into flats or pots. Most seeds will germinate within two weeks.

Potential Drawbacks:

The raw beans have potent trypsin inhibitor activity similar to that of soybean. The heat of cooking, however, destroys this property. The older tubers often take on an astringent taste, become very fibrous, and

Figure 21: *T. esculentum*

are difficult to consume. Also, the agronomic potential of Marama Bean is not very well studied, meaning there is very little available information relating to yield potential under specific conditions.

The plants do not tolerate waterlogged soils, and grow almost exclusively on sandy, nutrient-poor soil. Unlike most legumes, Marama Bean does not fix atmospheric nitrogen, and must rely on the available soil nitrogen, which is usually rather scarce. Therefore, the plant has evolved to become an efficient nutrient accumulator, which uses its tuber as a nutrient storage organ.

Nutrition:

The seed has an essential amino acid profile similar to that of soybean, and has more albumin and less globulin than soybean protein, which makes it more digestible by the human body.

Seeds (per 100g):	
Water	3.9g
Energy	635kcal
Protein	31.8g
Fat	42.2g
Carbohydrates	18.9g

Fatty Acid Composition:	
Oleic Acid	48-49%
Linoleic Acid	19-26%
Palmitic Acid	12-14%
Stearic Acid	7-10%
Arachidic Acid	3%

Essential Amino Acid Composition	(per 100g):
Tryptophan	219mg
Lysine	1119mg
Methionine	257mg
Phenylalanine	874mg
Threonine	822mg
Valine	1149mg
Leucine	1774mg
Isoleucine	1119mg

5-Month-Old Tubers	(per 100g):
Water	92.1g
Protein	2.1g
Fat	0.1g
Carbohydrates	4.4g

Source: PROTA
(Plant Resources of Tropical Africa)

Related Species:

Tylosema fassoglense shares a similar habitat as *T. esculentum*, but grows as far north as Sudan. It is used in the same ways as *T. esculentum* and produces seeds with a fat content greater than 30% and a protein content greater than 40%.

Moth Bean

Scientific name: *Vigna aconitifolia*
Family: Fabaceae
Synonyms: Moth Gram, Mat Bean, Matki Bean, Mout Bean, Dew Gram, Dew Bean
Zone: All Zones (Can potentially be grown wherever summers are hot and long enough)

Overview:

Moth Bean (pronounced 'moat') is a low-growing, mat-like, annual groundcover, which only grows up to 40cm tall, with runners up to 1.5m (5ft) in length. The plant grows during the summer months and matures from seed in 75-90 days. Moth Bean is self-fertile, and produces beautiful yellow flowers that turn into 2.5-5cm long pods containing 4-9 small seeds. The plant is valued as a vegetable and staple crop.

Habitat and Requirements:

Moth Bean is native to western - northwestern India, and into Pakistan. The largest concentration of the plants is found in India's driest state, Rajasthan. This is an area that receives the majority (85%) of

Figure 22: *V. aconitifolia* seeds.

its annual rainfall in the form of erratic summer monsoons from July-September. Here, daytime summer temperatures regularly reach 45°C (113°F), proving that Moth Bean tolerates extreme heat well. Moth bean grows well in areas that receive between 200-750mm of annual rainfall, but can also successfully complete its life cycle with as little as 50mm of rainfall. In rainfed systems a 1:1 ratio of early to late-producing genotypes should be planted, as erratic rains may cause one genotype to fail.

The plant requires full sun, and prefers to grow in sandy or sandy loam soil. Nonetheless, Moth Bean plants tolerate most soil types as long as they are not subjected to waterlogging. Excess moisture can easily lead to root rot and other fungal problems. The plant is moderately salt-tolerant, and has an extremely wide pH tolerance, which allows it to grow in soils that have a pH of 3.5-10.

Uses:

Fruit and Seeds:

The whole pods are eaten as a vegetable while still green. More commonly, however, just the seeds are sought-after from the ripe pods. Under optimal, controlled conditions Moth Beans have produced up to 2,600kg of seeds per hectare. Usually, however, they only produce 70-270kg/ha. Other sources suggest that as much as 1,800kg/ha can be produced. The pods and seeds can also be fed to livestock.
Leaves:
The foliage is a choice fodder for livestock, and plants commonly produce vegetative yields of 37-50t/ha. Moth Bean foliage can also be used for hay, and has roughly the same nutritional value as alfalfa hay.
Miscellaneous uses:
Due to their mat-like growth, the plants act as living mulch, which prevents the soil from eroding. Furthermore, the plant is nitrogen-fixing, and is able to colonize nutrient-poor soils. Moth Bean should be inoculated with the *Bradyrhizobium* strains that are used as Cowpea (*Vigna unguiculata*) inoculants.

Figure 23: *V. aconitifolia* foliage.
Credit: Sharad P. Nene

Propagation:

Moth Bean is propagated by seed. Seeds are commonly direct-sown at the beginning of the rainy season into cleared soil at a depth of 2.5-4cm. Optimum temperature for germination is 25-27°C (77-80°F). Under ideal conditions, most seeds will germinate within two weeks.

Potential Drawbacks:

The seeds contain a trypsin inhibitor, which needs to be destroyed by cooking (although it has a much weaker inhibiting effect when compared with that of soybeans).

Nutrition:

Seeds (per 100g):	
Water	9.7g
Energy	343kcal
Protein	22.9g
Fat	1.6g
Carbohydrates	61.5g
Calcium	150mg
Magnesium	381mg
Phosphorus	489mg
Iron	10.9mg
Zinc	1.9mg
Vitamin A	32IU
Thiamin	0.56mg
Riboflavin	0.09mg
Niacin	2.8mg
Vitamin B6	0.37mg
Folate	649mcg
Ascorbic Acid	4mg

Fatty Acid Composition	(per 100g):
Linoleic Acid	485mg
Palmitic Acid	313mg
Linolenic Acid	265mg
Oleic Acid	129mg
Stearic Acid	51mg

Essential Amino Acid Composition	(per 100g):
Tryptophan	147mg
Lysine	1248mg
Methionine	220mg
Phenylalanine	1028mg
Valine	734mg
Leucine	1541mg
Isoleucine	1138mg

Source: PROTA
(Plant Resources of Tropical Africa)

Related Species:

Moth Bean belongs to the same subgenus as Mung Bean (*Vigna radiata*), Rice Bean (*V. umbellata*), Black Gram (*V. mungo*), and Adzuki Bean (*V. angularis*).

Bambara Groundnut

Scientific name: *Voandzeia subterranea*
Family: Fabaceae
Synonyms: *Vigna subterranea*; Bambaranut, Congo Goober, Earth Pea, Baffin Pea, Njugo Bean, Madagascar Groundnut, Ground Bean, Earth Bean, Earth Nut
Zone: All Zones (Can potentially be grown wherever summers are hot and long enough)

Overview:

The Bambara Groundnut is an herbaceous annual that grows as a sprawling groundcover, or as an upright, bunching plant. The plant is similar to the peanut, because its growth habit entails burying its ripening fruit underground. Although the seed is edible and nutritionally valuable, it has received much less research attention than the peanut. Bambara Groundnut plants are self-fertile, and usually self-pollinated. The leaves of Bambara groundnut are trifoliate, and grow from the base of the plant on long stems. Flowers are yellow, and, when pollinated, send down a tendril into the soil, which eventually forms a pod. Pods usually contain only one seed. The seeds, with various earth tone colors, are between 1-1.5cm in diameter.

Habitat and Requirements:

The wild form of Bambara Groundnut is believed to be native to northeastern Nigeria and northern Cameroon. The cultivated form, on the other hand, is found growing in nearly all the African countries between the Tropic of Cancer and the Tropic of Capricorn. The plant has also managed to become pantropical, and is found in Asia and South America. The domesticated form, the subject of this entry, grows without help in areas that receive 300-3000mm of average annual rainfall. It grows optimally, however, in areas that receive 600-1000mm of annual rainfall. The plant grows in areas that experience daytime maximum temperatures a little over 40°C (104°F), but it performs optimally in regions that have temperatures between 20-30°C (68-86°F).

The plant grows best in areas of sandy loam soil, but performs well in most well-drained, nutrient-poor soils. The plant tolerates a soil pH range of 4.3-7, but grows optimally wherever soil pH is between 5 and 6.5. It is resistant to soil aluminum, growing where most other crops would fail.

Uses:

Fruit and Seeds:

Pods are usually ready to be harvested in 130-150 days after germination. Generally, the pods are harvested when the plants turn yellow. At this stage, they are fully mature, and about as hard as black beans. They can be roasted, ground into flour or boiled and eaten. Similar to black beans, they take a long time to soften when cooked. The seeds can also be harvested before the plants turn yellow, and, while still green, the immature seeds are boiled with salt and eaten as a snack. The seed oil can be used for cooking, but only makes up about 6% of the seeds. The plant usually produces 300-600kg of seeds per hectare, but under ideal conditions the plants can yield up to 4 tons/ha.

Leaves:

The foliage (as well as the seeds and pods) can be used as fodder for livestock, and is high in protein and phosphorus.

Figure 24: *V. subterranea.*

Miscellaneous uses:
The plants are nitrogen-fixing, and nodulate best with *Bradyrhizobium* strains. Therefore, the plants can be used to reclaim degraded soils, as cover crops, and as companion plants for other crops.

Propagation:
The seeds are best stored at freezing temperatures, but even at room temperature, they have a long viability as long as they are not attacked by pests. The seeds should be removed from their pods, and soaked for 12 hours in water prior to sowing. The seeds are best sown direct into well prepared, loosened soil at a depth of 5cm. Most seeds will germinate in 7-21 days.

Potential Drawbacks:
It is often necessary to mound soil around the plants, because the fruits may fail to bury themselves deeply enough. Also, the seeds contain tannins and low levels of trypsin inhibitors, which appear to be

relatively heat stable. In any case, the seeds have a long history of safe consumption. Cooking, soaking, sprouting, fermenting, and other forms of processing help reduce the anti-nutritional factors.

Nutrition:

Raw Immature Seeds	(per 100g):
Water	57.3g
Energy	152kcal
Protein	7.8g
Fat	3.1g
Carbohydrates	30.0g
Fiber	3.0g
Ash	1.8g
Calcium	14mg
Phosphorus	258mg
Iron	1.2mg

Mature Dry Seeds	(per 100g):
Water	10.3g
Energy	367kcal
Protein	18.8g
Fat	6.2g
Carbohydrates	61.3g
Fiber	4.8g
Ash	3.4g
Calcium	62mg
Phosphorus	276mg
Iron	12.2mg
Beta-carotene	10mcg
Thiamin	0.47mg
Riboflavin	0.14mg
Niacin	1.8mg

Essential Amino Acid Composition	(per 100g):
Tryptophan	192mg
Lysine	1141mg
Methionine	312mg
Phenylalanine	991mg
Threonine	617mg
Valine	937mg
Leucine	1385mg
Isoleucine	776mg

Fatty Acid Composition of Seed Oil:	
Palmitic Acid	18-24%
Stearic Acid	5-12%
Oleic Acid	18-24%
Linoleic Acid	34-40%
Linolenic Acid	2-3%
Behenic Acid	3-7%

Dried Leaves	
Crude Protein	15.9%
Crude Fiber	31.7%
Ash	7.5%
Fat	1.8%

Source: PROTA
(Plant Resources of Tropical Africa)

Related Species:

Macrotyloma geocarpum is another herbaceous annual with a similar growth habit that is much lesser know than *Voandzeia subterranea. M. geocarpum*, also known as Kersting's Groundnut, is more drought-tolerant than Bambara Groundnut. Kersting's Groundnut produces an edible seed with 19-20% protein, of which 6.2% of the total essential amino acid content is made up of lysine, and another 1.4% is composed of methionine. Another similar species deserving further investigation is *Voandzeia poissoni.*

Part 2: Palms

The classic desert oasis tree is a palm. Few plants are as reliably beautiful as the palms. Furthermore, in many parts of the world, people depend on palms in order to maintain a basic quality of living. People often harvest parts of palm trees as construction materials, to make craft utensils, and even to fashion articles of clothing. The palms presented in this book are often used for such purposes, and have the added benefit of providing a myriad of edible yields.

Jelly Bean Palm

Scientific name: *Butia capitata*
Family: Arecaceae
Synonyms: Pindo palm, Jelly palm
Zone: 8b

Figure 25: *B. capitata*. Santa Barbara, California.
Credit: Mark Viergutz

Overview:

The Jelly Bean Palm is a slow-growing, shade-tolerant, monoecious feather palm up to 7m (23ft) tall. It has a trunk diameter about half a meter wide and a crown spread of up to 5m (16ft). The beautiful, blue-green fronds incurve towards the trunk. Whitish flowers give rise to plump, bright orange-yellow fruits, which are variable in taste and up to 3cm wide. The fruits can range from fibrous to juicy, and from tart to sweet and usually ripen in late fall. The trunk of the palm is known for its old leaf bases that remain attached for years.

Habitat and Requirements:

Jelly Bean Palms are native to the area encompassing southern Brazil, northeastern Argentina, and northern Uruguay. They grow in part-shade and full sun, and tolerate a wide range of soils, including soils that are slightly alkaline or acidic. They tolerate sand and clay soil.

Pindo Palms also tolerate a wide range of climatic conditions. For example, they grow equally well in the hot and humid deep-south as they do on the Mediterranean west coast of the United States. They can grow well in coastal Mediterranean climates that receive as little as 300mm of annual precipitation provided that they receive some summer shade and are not planted in sandy soil.

Figure 26: Pindo Dates ready for the market

Uses:

Fruit and Seeds:

The ripe fruits are juicy and delicious. The flesh can be eaten raw, made into preserves, or juiced and tastes like a cross between pineapple and apricot. The seed shells can be cracked open and the seeds eaten. The seeds taste like a miniature coconut, and are sometimes referred to as "coquitos." There are many species of palms that yield coquitos.

Bark:

Since the old leaf stalks remain on the tree for years, the trunks of Pindo Palms can be used as a vertical growing space for small plants. Indeed, bird feces often land between the leaf stalks and trunks where the humid microclimate germinates seeds. Many plants, including figs and columnar cacti, have been seen growing from these "pockets." One can fill the pockets with a rich potting soil and propagate small, shade-tolerant vegetables such as Okinawan Spinach (*Gynura crepioides*) in order to maximize production in a small amount of space.

Miscellaneous uses:
The palm hearts can be eaten, but harvesting will kill the palm unless there are branches or suckers, which is rare for this species. Since the Jelly Bean Palm is so slow-growing, harvesting of palm hearts could easily lead to the population's decline.

Propagation:

Like most palm species, Pindo Palm seeds can be propagated with the "baggy method," which involves placing seeds and moist peat moss in air tight plastic bags. Alternatively, seeds can be propagated in storage totes by poking drainage holes in the storage tote and filling the bottom with a 50/50 mix of moist perlite and vermiculite. The lid should be lifted to check for germination once per month. The seedlings are easy to transplant out of this mixture.

Potential Drawbacks:

The variability of fruit quality from one palm to the next is an issue without an easy resolution. Since the Jelly Bean Palm does not usually produce suckers, it is extremely difficult to establish reliable cultivars using conventional methods.

Nutrition:

Jelly Bean Palm fruits or "Pindo Dates" are a good source of provitamin A and Beta-carotene. The fruits may also be effective as a vermifuge.

Related Species:

Taxonomical debate has been ongoing to determine the difference between *Butia capitata* and *Butia odorata*, or whether or not they are the same species, or whether or not the plant commonly known (and described here) as *Butia capitata* is actually *B. odorata*. *Butia eriospatha* is a very similar species, also with edible fruit.

Gingerbread Palm

Scientific name: *Hyphaene thebaica*
Family: Arecaceae
Synonyms: Doum Palm, African Doum Palm, Egyptian Doum Palm
Zone: 9b

Overview:

The Gingerbread Palm, a historically important palm, was apparently esteemed by the rulers of ancient Egypt such as King Tut who was discovered 3,000 years after his death with baskets of Gingerbread Palm fruits in his tomb. The Gingerbread Palm is a unique fan palm, which has the habit of branching. Its branching habit is so great that mature specimens have a wide, dome-like canopy that provides dappled shade to the ground below. The Gingerbread Palm can grow up to a maximum height of 20m (66ft) with an equal spread.

Gingerbread Palms are dioecious. Female trees are highly productive and individual specimens can produce about 50kg of fruit per year. Under ideal conditions, trees begin to bear fruit after 6-10 years and remain in optimal production for about 60 years. The fruits take from 240 days to as long as one year to ripen. The fruits are edible, and the tree is exploited for many other uses as well.

Habitat and Requirements:

The Gingerbread Palm is native to the Nile River Valley in the modern day countries of Sudan and Egypt, and has since been widely cultivated throughout the Sahel of northern Africa. It grows well in areas that receive anywhere from 50-600mm of average annual precipitation. Gingerbread Palms indicate that there is a high water table, and are often found near seasonal or perennial bodies of water. They are common along the upper banks of the Nile River.

The Gingerbread Palm tolerates and thrives in extreme heat as long as there is ample groundwater. It is not a very cold tolerant palm, and should only be considered hardy down to 28°F (-2°C). Gingerbread Palms require full sun.

Gingerbread Palms require very well-drained soil, and are almost always found growing on sandy sites. Upon germination, they concen-

Figure 27: *H. thebaica*. Porga, Benin.
Credit: Marco Schmidt

trate most of their energy into growing a sinker root. This deep taproot easily moves through the sandy soil and stops once it has reached the water table. The palms are known to withstand brush fires, moderate salinity, and prefer soils with a pH of 6.5-7.5.

Uses:

Fruit and Seeds:

The mesocarp of the ripe fruit is edible, but can be very hard. It has a sweet, gingerbread-like taste and is often ground into a meal or made into syrup. The endosperm of unripe fruits is also edible and is similar to a coconut with sweet liquid inside. Endosperms from unripe fruits can be consumed raw or boiled. When mature, the endocarps of the fruits are very hard and can be used as a sort of vegetable ivory. The endocarp is often used as a small cup or container. A black dye can also be obtained from the dried, mature fruits.

Flowers:

The male inflorescence is often used as fodder for livestock.

Leaves:

Young leaves are cut as fodder for livestock before they become too unpalatable. More often, however, leaves are used for thatching and weaving. The unexpanded, young leaves are the most sought after for weaving baskets, bowls, hats, fans, bags, etc. Ropes are also commonly crafted using the leaves. The expanded leaves are most often used for roof thatching. The leaf petioles can be used to build fences, bridges, and are also woven into bed mats.

Wood:

The trunks can be used for general construction for structures, and can be hollowed out to be used as drain pipes, or cut into panels to make canoes. The wood can also be utilized as fuel, and in charcoal making.

Roots:

The roots yield a fiber that is useful in making fishing nets and animal snares.

Miscellaneous uses:

Ash from the trunk can be used as a salt substitute. When trees are close to flowering, their trunks are tapped for their sap, which is then fermented into palm wine. The stems of seedlings can be cooked and eaten as a vegetable, as well as the apical bud at the tip of the trunk, and the palm heart located in the center of the trunk towards the base of the tree. Since Gingerbread Palms branch, harvesting the apical bud can be done on individual branches without killing an entire tree. Tapping individual branches often only kills that branch, and in many cases, trees can be coppiced, and will resprout from their base.

Propagation:

Seeds should be direct sown. Due to their sinker root, nursery propagation is not very feasible and often leads to complications when transplanting. Like many palm species, germination can take a very long time (up to one year or more), but often occurs as quickly as one month. Both the pericarp and the endocarp, or just the pericarp can be removed to hasten germination. The best method to germinate the seeds, however, is to mechanically scarify them and then soak them in water for three days before planting. Seeds can also be placed in small mounds of mulch, the decomposition of which helps spur germination. There appears to be a significant loss of viability in the seeds over the course of as little as one year. Fresh seeds should be used for the greatest success.

Potential Drawbacks:

This species dislikes cold, wet, heavy soil. The wood of female palms is susceptible to attack from termites and other boring insects.

Nutrition:

Ground pericarp flour	(per 100 g):
Moisture	10.7g
Energy	296kcal
Protein	2.6g
Fat	.4g
Carbohydrates (including fiber)	79g
Fiber	14g
Ash	7.3g
Calcium	68mg
Thiamine	.05mg
Riboflavin	.1mg
Niacin	3.4mg

Kernel	(per 100 g):
Moisture	5.7-6.2g
Energy	395 kcal
Protein	2.4-5.0g
Fat	4.9-8.0g
Carbohydrates (including fiber)	80.6g
Fiber	6.5-11.0g
Ash	1.9-5.4g
Calcium	121-168mg
Phosphorous	170-281mg

Source: PROTA
(Plant Resources of Tropical Africa)

Related Species:

Hyphaene petersiana is the southern counterpart of the Gingerbread Palm. It is native from northwestern Namibia to the east through Zimbabwe. The palm is from a nearly identical climate in the south of the African content as the habitat of *H. thebaica.* Its appearance is much different, growing straighter and taller, often with a single trunk, and usually does not branch. On the other hand, *H. petersiana* is utilized in nearly identical ways as *H. thebaica.*

Chilean Wine Palm

Scientific name: *Jubaea chilensis*
Family: Arecaceae
Synonyms: Coquito Palm, Palma Chilena, Honey Palm, Wine Palm
Zone: 8a

Overview:

The Chilean Wine Palm is known for having the thickest trunk of any palm species. Individual specimens can have a trunk that is about 2m (6.5ft) in diameter. Chilean Wine Palms are cold-hardy feather palms that can grow up to heights of 25m (82ft) and have a crown spread of up to 7m (23ft). Individual fronds are pinnate, can grow up to 4m in length and often have a silvery, bluish tint. Historically, the palm

Figure 28: *J. chilensis*. Goleta, California
Credit: Mark Viergutz

has been felled and tapped for its sap, which can be fermented into wine. The plant produces yellow-orange, edible fruit that contain a seed with an edible endosperm.

It takes a long time for a tree to reach adult size. In some cases, it can take up to 20 years for an individual to develop a full-size trunk. The palm makes up for this slow growth, however, in terms of longevity. It is not uncommon for the Chilean Wine Palm to live for 400 years.

Figure 29: La Campana National Park, Chile
Credit: Alain de Trenqualye

Habitat and Requirements:

Chilean Wine Palms are native to the Mediterranean climate region of Chile between 31-35° south latitude (about the same distance from the equator as southern California). Winters are rainy and cool, and are followed by a long dry season that lasts throughout the majority of the year and is concentrated in the summer. The average annual rainfall throughout the Wine Palm's natural range is between 300-700mm.

The palms are found in a diversity of locations, including upland rocky areas and alluvial lowlands. The palms are found growing up to an elevation of about 1,370 meters (4500ft.). Wine palms prefer well-draining soil, but grow well in clay loams, and even grow moderately in heavy clay. They are not adaptable to humid regions. When young they prefer partial shade, but as they mature they require full sun.

Uses:

Fruit and Seeds:

The fruits turn orange or yellow when ripe and fall from the tree. The fruits are not very juicy, so they tend to take a long time to rot on the ground after falling from the tree. The fruit is mildly sweet and has a refreshing taste that does not overwhelm the palate. The texture is similar to a Rose Apple (*Syzygium jambos*). The seeds can be cracked open to reveal a sweet-tasting endosperm that tastes almost identical to coconut flesh. These seed endosperms are often sold in markets in Latin America with the name, "coquitos."

Leaves:

The main stem of the leaves can be used for fencing. The fibers can also be used as stuffing for furniture.

Miscellaneous uses:

Older trees can be felled and their crowns removed in order to collect the sweet sap, which is then boiled down to produce what is known as "miel de palma." The sap that exudes from the top of the fallen trunk has a high sugar content, and can be continually harvested over the course of several months as long as small slices of trunk are cut off every day. Depending on the size of the palm, a felled trunk can yield up to 340 liters (90 gallons) of crude sap. Felling is best executed in October (April in the Northern Hemisphere) and usually lasts until the end of April (October in the Northern Hemisphere).

Propagation:

Chilean Wine Palms are propagated from seed. They are simple to propagate and work with in the nursery, but, like many palms, take a long time to germinate. Seeds can take anywhere from two months to two years to germinate, but generally germinate towards the lower end of that range. No pretreatment is necessary. One should use a plastic storage bin with drainage holes drilled in the bottom, and the bin should be filled with a 50/50 mixture of moist perlite and moist vermiculite. The seeds are then placed in the mixture, buried slightly, and irrigated. The lids should be lifted once per month to irrigate, and to see if there are any seedlings that have germinated.

There is no harm in leaving seedlings in the storage bin for a couple of months, but it is best to transplant the seedlings soon after they germinate, before they develop a large root. The seedlings are then dug out of the mixture and potted up in the nursery. Since Chilean Wine Palms often germinate in the understory of shrubs and other plants, it is important to raise the seedlings in diffused sunlight/light shade for their first 1-2 years.

Nutrition:

Seed endosperm ("Coquito"):

Nutrition Facts (Serving size 11 coquitos)	
Calories	110
Calories from Fat	90
Total Fat	10g
Saturated Fat	9g
Sodium	5mg
Total Carbohydrates	5g
Fiber	3g
Sugars	1g
Protein	1g

Source: Frieda's The Specialty Produce People

Related Species:

Jubaea chilensis is the only member of its genus.

Mountain Coconut

Scientific name: *Parajubaea torallyi var. torallyi*
Family: Arecaceae
Synonyms: Bolivian Mountain Coconut, Janchicoco, Pasopaya Palm
Zone: 9a

Overview:

Mountain Coconut palms grow quickly compared to most palms, and eventually attain heights of up to 14 meters (46 feet). They develop thick trunks that nearly rival the girth of Chilean Wine Palm trunks. Individual fronds grow to about 5m (16ft) in length. The foliage is very ornamental, and the trees are often planted as a substitute for coconuts in climates where coconuts cannot grow. The trees are monoecious, and therefore require no cross-pollination to set fruit. The fruits are large and woody, but much smaller than, and not nearly as economically important as the coconut. The tree starts producing fruits after 10-15 years.

Habitat and Requirements:

The Mountain Coconut is native to a small number of mountainous valleys and ravines located to the east of Sucre, Bolivia. The parent material on which the palms grow is predominantly sandstone. The palms can be found growing up to an altitude of 3,400 meters (11,200 feet). The climate of the Mountain Coconut's native habitat is rather arid. It is not uncommon for the region to experience droughts where no rain falls for ten months. In general the region receives its 2-4 month rainy season during the summer months.

Although the plant is endangered in the wild, it has been cultivated extensively by palm enthusiasts in recent years. Mountain Coconuts seem well-adapted to coastal California and parts of Australia, suggesting that they not only grow well in arid and semi-arid, subtropical highlands, but also in Mediterranean climates, and other mild, dry climates. Despite what their native habitat suggests, the palms do not require a rainy season during the summer in order to successfully adapt outside of their native range. The palms adapt well to areas that receive roughly 300-900mm of average annual rainfall.

Figure 30: A young *P. torallyi var. torallyi*. Santa Barbara, California

The palms prefer well-drained soil, and dislike waterlogged soils, but will tolerate clay soil as long as it has been amended to some degree before planting. The palms seem to grow best when daytime temperatures are between 16-24°C (60-75°F). The palms grow quickly compared to most palms, and sometimes tolerate very hot temperatures, but will slow their growth rate during heat waves. Prolonged, extreme heat may prove fatal. Mountain Coconuts require full sun in order to grow well. Palms may survive temperatures down to -8°C (17°F), but will be damaged once temperatures drop into the low 20s Fahrenheit.

Uses:

Fruit and Seeds:
The thick, woody shells of the fruits are cracked open, and the oil-rich kernels eaten. The shells are extremely difficult to crack. The fruits can also be fed to animals.

Leaves:
The fibers of the fronds are woven in order to create a variety of useful objects such as floor mats, baskets, and ropes. The leaves of young plants are eaten by livestock. The leaves from mature trees can also probably be fed to livestock. The midrib of mature leaves can be burned as a fuel source, and the stalks can be used for construction.

Miscellaneous uses:
The trees can be harvested for their edible palm hearts. The sap can also be tapped and used to make a refreshing drink.

Propagation:

Mountain Coconuts are difficult to propagate, because the trees are not vegetatively propagated and seeds erratically germinate. Seeds can take anywhere from a few months to a few years to germinate. Part of the pericarp can be cut in order to allow humidity to more easily pass through to the center of the seed. Additionally, the seeds can be soaked in water or a solution of Gibberellic Acid for a week before planting. The seeds can be placed in plastic bags in moist peat moss. The seeds can also be buried slightly in a sealable container filled with 50% perlite and 50% vermiculite, and kept moist. Mountain Coconuts tolerate container culture well.

Potential Drawbacks:

The woody shells of the fruits are thick and require strong tools in order to be efficiently cracked.

Nutrition:

The nutritional profile is probably similar to that of the "coquitos" produced by Chilean Wine Palms (*Jubaea chilensis*).

Related Species:

Parajubaea coccoides is a palm that grows in the Andes Mountains of Ecuador in and around the city of Quito. As such, this species is not as drought-tolerant as *P. torallyi var. torallyi.* This palm produces good-tasting fruit.

Parajubaea sunkha is native to the province of Santa Cruz, Bolivia. *P. sunkha* is the smallest species of *Parajubaea.*

Aside from *P. torallyi var. torallyi*, the only other variety of *Parajubaea torallyi* is *Parajubaea torallyi var. microcarpa.* This variety is smaller and produces smaller fruit that are said to be better-tasting.

Date Palm

Scientific name: *Phoenix dactylifera*
Family: Arecaceae
Zone: 8b

Overview:

The Date Palm is an elegant palm that can attain a maximum height of up to 36m (118ft). The palms are dioecious, and commercial orchards usually plant one male tree for every 50 female trees. The pinnate, bluish-green fronds can grow up to 6m (20ft) in length and have sharp spines at their base. The plant begins flowering when it is 6-10 years old. In the wild, palms are pollinated by the wind, but under cultivation they are usually hand pollinated. Depending on the cultivar, an individual female tree produces between 25-140kg (50-300lbs) of fruit per year.

Habitat and Requirements:

The Date Palm is believed to be native to parts of the Middle East surrounding the Persian Gulf where average annual rainfall is about 100-300mm. The plant prefers to receive most of its rain during the winter, because rainfall during the summer and fall (when the plant flowers and sets fruit) will often cause crop failure due to fungal diseases. Therefore, Date Palms require low humidity. The Date Palms are largely riparian, and grow in areas where there is ample groundwater and a high water table. It is the classic desert oasis tree.

The palms grow where drainage is good and tolerate sandy to clay soils as long as the clay drains well. They tolerate alkalinity very well, and are moderately tolerant of salinity.

The palms tolerate greywater and other forms of reclaimed water very well. They can be planted close together in a circle around a basin for greywater, and will grow away from each other. They are very productive when planted in this manner. The plants tolerate shade when they are young, because they often grow naturally in the understory of other plants. However, when mature, they require full sun in order to produce well.

Uses:

Fruit and Seeds:

The fruits have an excellent taste, and can be eaten fresh, but are most often pitted and dried for later use. Non-pollinated fruits can be used as feed for animals. The seeds can also be ground up and fed to many animals including chickens, horses, camels, cattle, sheep and goats. Seeds can also be roasted and ground as a coffee substitute or mixed with flour to make very poor-quality bread.

Flowers:

The inflorescences can be tapped for sap, which is then used to produce sugar. Tapping the trees hampers fruit production, and should only be done sparingly.

Leaves:

The young leaves and terminal bud (heart) can be cooked and eaten as vegetables. Removal of the heart kills the palm, and a thick, sweet liquid pools in its place, which can be drunk, but quickly ferments into an intoxicating beverage. The leaves can also be used to make a variety of crafts and utensils such as hats, mats, screens, fans, and

baskets.
Wood:
The trunks can be used for general construction wherever posts or columns are needed, but are not very strong, and are better used as fuelwood.

Figure 31: *P. dactylifera* oasis. Misfat Al Abriyyen, Sultanate of Oman
Credit: Cathy Birdsong Dutchak

Propagation:
Propagation of date palms can be done from seed using the same methods to propagate other palms. The palms, however, do not come true to seed, and therefore seed propagation is largely irrelevant in commercial Date operations unless a grower is consciously trying to expand the genetic base, and breed new varieties. Plants must be asexually propagated in order to maintain cultivars.

Traditionally, the plants are propagated via removal of the trunk offshoots from the parent plant. Offshoots are best removed when they are 3-5 years old. The soil around the parent plant should be deeply irrigated a couple days before removing the offshoot in order to reduce the stress of removal. Most of the soil is dug away from the parent plant to reveal the union point where the trunk of the parent tree and offshoot meet. Then a sharp, metallic chisel is used to slowly cut the offshoot from the parent trunk. The offshoot should be left to callous in the shade for 10-15 days. The offsets should be raised in the nursery for the next one to two years before being planted out in the field. The survival of the offshoots depends largely on the variety of palm. Medjool's offshoots are notoriously more difficult to root compared to the easier, Deglet Noor and Zahidi.

More recently, the plants have been propagated by tissue culture, which requires sterile laboratory conditions.

Potential Drawbacks:

Since many cultivars send out a maximum of 20-30 offshoots in their lifetime, individual palms can turn into clumps. If the offshoots are not removed, the palms will put more energy into the vegetative growth of the offshoots instead of their flowers and fruits. Single-trunked palms should be maintained for optimum fruit production.

Offshoots are produced starting when the tree is about 10-15 years old. Mektoum and Barhee are two varieties that produce very few offshoots.

Nutrition:

Nutrient data for Deglet Noor Dates (per 100g):

Water	20.53g
Energy	282kcal
Protein	2.45g
Total lipid (fat)	0.39g
Carbohydrates	75.03g
Fiber	8.0g
Sugars, total	63.35g

Calcium	39mg
Iron	1.02mg
Magnesium	43mg
Phosphorus	62mg
Potassium	656mg
Sodium	2mg
Zinc	0.29mg

Vitamin C, total ascorbic acid	0.4mg
Thiamin	0.052mg
Riboflavin	0.066mg
Niacin	1.274mg
Vitamin B-6	0.165
Folate, DFE	19mcg_DFE
Vitamin B-12	0mcg
Vitamin A, RAE	0mcg_RAE
Vitamin A, IU	10 IU
Vitamin E (alpha-tocopherol)	0.05mg
Vitamin K (phylloquinone)	2.7mcg
Fatty acids, total saturated	0.032g
Fatty acids, total monounsaturated	0.036g
Fatty acids, total polyunsaturated	0.019g

Source: USDA National Nutrient Database

Related Species:

Phoenix canariensis, a native of the Canary Islands, is well-adapted to Mediterranean climates. The fruits barely have any flesh, but are sweet. The trunk is tapped for its sweet sap, which can be condensed into a "palm honey" that tastes similar to maple syrup. This palm does not require groundwater, and often succeeds as an upland species in very dry areas.

Phoenix reclinata, known as the Senegal Date Palm, produces edible fruits, which are also inferior to those of *P. dactylifera*. The seeds are roasted and used as a coffee substitute, and the inflorescence can be tapped to make sugar. This palm grows in similar habitats as *P. dactylifera*.

Licuri Palm

Scientific name: *Syagrus coronata*
Family: Arecaceae
Synonyms: Ouricuri Palm, Aricuri, Nicuri, Baba-De-Boi
Zone: 9b

Overview:

The Licuri Palm is a single-trunk feather palm growing up to 12m (39ft) tall. This monoecious palm is very ornamentally pleasing as its leaf bases remain on the trunk in a spiral pattern after the leaves have fallen. The plant is commonly known as the primary fodder species for the Leer's Macaw, which feeds on the endosperm of the seeds. The palm has a myriad of other uses, and is an economically important species in its native range.

Figure 32: *S. coronata*. Euclides da Cunha, Brazil
Credit: Marcelino Ribeiro

Habitat and Requirements:

The Licuri Palm is native to the thorny scrub biome of northeastern Brazil known as the Caatinga, as well as the surrounding tropical savanna. The palm grows well in areas that receive 400-1000mm of average annual rainfall. The majority of this rainfall occurs during the summer months, and in the Caatinga the plant usually experiences a dry season that lasts for 7-11 months. In this region, Licuri Palms are commonly found growing alongside cacti. Historically, in the Caatinga region, rainless droughts have occurred that lasted for several years.

The palms prefer growing on sandy, well-drained soil, but also tolerate clay soil and even withstand heavy clay soil during the cool, wet winters of Mediterranean climate regions. Licuri Palms tolerate prolonged periods of extreme heat. They grow well in partial shade when young, but prefer full sun as they mature.

Uses:

Fruit and Seeds:

The fruits are edible raw or cooked and have a sweet flavor and yellow-orange color when ripe. The pulp, however, is not nearly as sought after as the seeds, which contain a white, oily endosperm surrounded by a hard endocarp. The endosperm of the Licuri Palm has the richest flavor of all the "coquitos" mentioned in this book. The oil from the endosperm is also commonly extracted for use in the manufacturing of soap. The entire fruits can also be fed to chickens or livestock.

Flowers:

The inflorescence can be used as fodder for livestock.

Leaves:

The leaves can be used as fodder for livestock. The wax taken from the leaves is used to manufacture soaps. The petioles of the leaves are used to make bird cages or small animal hutches. Additionally, the leaves are woven into baskets, made into brooms, and used to make strings, hammocks and roofs for houses. Most commonly, however, the leaves are made into large, straw sunhats.

Wood:

The trunks are sometimes used for construction.

Miscellaneous uses:

The debris left over from kernel extraction is burned as fuel or made into charcoal. The starchy pith of the tree can be made into a flour substitute.

Propagation:

The seeds should be treated like those of other palm species, and buried halfway into a 50/50 mixture of moist perlite and vermiculite in large, plastic storage bins with drainage holes. Otherwise, the seeds can be individually sown into deep-rooted, narrow containers. The seedlings develop a thick taproot, which can be bare-rooted briefly when transplanting. If direct seeding, the seeds can also be placed in the middle of small mounds of mulch/compost. Licuri Palms tolerate container culture very well, and can be left in the nursery for many years.

Nutrition:

Fruit Pulp:	
Moisture	77.4%
Calorific value (per 100g)	108.6kcal
Ash	1.4%
Lipids	4.5%
Nitrogen	0.5%
Protein	3.2%
Total Carbohydrates	13.2%
Beta-Carotene (mcg/g)	26.1
Vitamin A (mcg/g)	4.4
Alpha-Tocopherol (mcg/g)	3.8

Nut (seed endosperm):	
Moisture	28.6%
Calorific value (per 100g)	527.3kcal
Ash	1.2%
Lipids	49.2%
Nitrogen	2.2%
Protein	11.5%
Total Carbohydrates	9.7%

Source: Composição nutricional do fruto de licuri

Related Species:

Syagrus vagans has a very similar distribution range, and is used in much the same ways as *S. coronata*.

Part 3: Cacti and Succulents

Of all the plants in the plant kingdom, there are none that better represent a dryland setting than cacti and succulents. All cacti are succulents, but not all succulents are cacti. Both have thick skin and are well adapted to surviving very dry, harsh conditions. Cacti are especially well adapted to dry areas with their extensive root systems that suck up the moisture from brief storms. They also come equipped with thick skin and modified leaves (spines), efficiently reducing water loss through evapotranspiration. All cacti have edible fruit (although the quality of many species is not very good). Some species even have edible stems that can be used as a vegetable. The species of cacti presented in this book have noteworthy fruit, and some have edible stems. The succulents presented in this book display important attributes, which may contribute to their utilization as survival foods, or to their use in more valuable economic ventures.

Aloe Vera

Scientific names: *Aloe vera*, *Aloe barbadensis*, *Aloe indica*
Family: Aloeaceae
Synonyms: Chinese Aloe, True Aloe, Burn Aloe
Zone 8b

Overview:

Few plants are as commonly grown and forgiving of neglect as *Aloe vera*. This common succulent is grown as a house plant in many parts of the world, and has achieved the greatest commercialized status of any plant in this book.

Habitat and Requirements:

Little is known with exactitude regarding the origins of *Aloe vera*. Recently however, hypotheses based on DNA comparisons linking it to similar species, suggest that *Aloe vera* is native to the region now dominated by the Sahara desert. The loss of the dry sclerophyll forest encompassing large parts of the Sahara desert during more humid centuries, has contributed to the fracturing of wild populations of *Aloe vera* and its relatives. Common Aloe's most notable relatives are *Aloe perryi*, *A. forbesii*, *A. inermis*, *A. scobinifolia*, and *A. sinkatana*. These species are native to the region in and around Sudan, at the eastern end of the Sahara desert. It is thought that *Aloe vera* has descended from a hybridization of any combination of these species.

Although, descending from desert regions, *Aloe vera* performs equally well in tropical climates provided it has good drainage. The plants are salt-tolerant, and grow well in seaside areas. Actual annual average rainfall amounts tolerated by this species are between 200-2000mm. The higher end of this range requires excellent drainage to keep plants from rotting, while the lower end of this range requires some form of shade to keep the plants from desiccating. In Mediterranean climates with 300-600mm of annual rainfall plants perform very well in the understory of forests. In part shade the plants become much more robust, with a deeper green compared to plants grown in full sun.

Uses:

Fruits and Seeds:
The seeds contain edible oil similar to sunflower oil, and can be used for making biodiesel.
Flowers:
Flowers are borne at the end of a long stalk. The stalk can be pickled.
Leaves:
The cosmetic industry widely uses the gel from the inside of the leaves. Topically *Aloe vera* gel may be found in any product from sunscreen to makeup, shaving cream, anti-wrinkle cream and shampoos. Although there is little scientific proof to back many of the claims marketers use to sell their products (especially in the case of sunscreens), it has been proven that Aloe gel can be applied topically for many benefits. *Aloe vera* gel has antibacterial and antifungal properties, which gives it the ability to help cure less serious skin infections. These properties may also contribute to the fact that the gel helps heal first and second degree burns, although that effect is likely due to the gel's moisturizing qualities.

While topical applications of *Aloe vera* leaves have received little scientific research, internal properties have been studied extensively. *Aloe vera* has been shown to boost the immune systems of cats and dogs, helping them to fight cancer, suggesting that it probably serves the same function in humans. Before deciding to consume *Aloe vera*, one should know that unprocessed *Aloe vera* contains the compound Aloin, which has been proven to have a laxative effect. In the U.S. many companies remove this compound before adding Aloe to their beverages and desserts. Nonetheless, consuming one or two fresh *Aloe vera* leaves per day appears to have no noticeable laxative effect, suggesting that laxative effects are probably only achieved through the consumption of large amounts of Aloe or in sensitive people. In many parts of Latin America street vendors sell a juice made from Aloe, jugo de sabila, as a refreshing drink. These street vendors as well as many supermarkets around the world use fresh leaves, suggesting that negative effects resulting from fresh *Aloe vera* consumption are rare.

Conversely, *Aloe vera* has been shown to help digestive problems such as heartburn and irritable bowel syndrome. Consumption may also be helpful in treating eye-disease, tumors, spleen enlargement, liver complaints, vomiting, bronchitis, biliousness, asthma, jaundice and ulcers.

Figure 33: *A. vera*. Santa Barbara, California

Propagation:

Aloe vera is most commonly cultivated by removing the suckers that form around the parent plant. When separating from the parent plant care should be taken to preserve the roots of the suckers. Nonetheless, rootless suckers easily form roots once they are potted directly into containers with a well-draining potting mix in diffused light. Roots should form within two weeks even without rooting hormone.

Although much literature suggests that *Aloe vera* may be propagated from seed, there are also very few reports of sexual propagation. Since *Aloe vera* is somewhat of a taxonomical nightmare, reports of *Aloe vera* seed propagation are most likely the result of a hybridization between *Aloe vera* and another similar species. Due to the fact that it has been asexually propagated for so long, most *Aloe vera* plants in circulation are clones of the same distant parent plant.

Aloe vera appears to be self-sterile. The author has propagated two different batches of seed from cross-pollinated *Aloe vera* plants, and on both occasions the germination rate was zero percent. Since the health benefits of *Aloe vera* don't appear to be exclusive to that species, *A. vera* should be hybridized with other, similar plants of the same genus. The resulting hybrids could be used to select superior traits. Hybrid offspring are the most likely reason that people suggest *A. vera* can be propagated from seed.

Potential Drawbacks:

Without adequate drainage, plants are prone to rotting, especially in cold and wet soils.

Nutrition:

Amino acids: Isoleucine, leucine, lysine, methionine, phenylalanine, threonine, valine

Anthraquinones: Aloe-emodin, aloetic acid, aloin, anthracene, anthranol, barbaloin, chryscophanoic acid, emodin, ester of cinnamic acid, ethereal oil, isobarbaloin, resistannol Enzymes: Alkaline phosphatase, amylase, catalase, lipase, oxidase

Minerals: Calcium, chromium, copper, magnesium, manganese, sodium, zinc

Prostanoids: gamma-linolenic acid

Saccharides: Aldopentose, cellulose, glucose, L-rhamnose, mannose
Vitamins: Alpha-tocopherol, beta-carotene, choline, folic acid, vitamins B1, B2, B6, C, and E
Source: Plants for a Future Database

Related Species:

As mentioned already, *Aloe vera* is closely related to many other *Aloe* species with similar properties. These species include *Aloe perryi, A. africana, A. spicata, A. forbesii, A. inermis, A. scobinifolia, A. sinkatana*, and *A. striata*.

Apple Cactus

Scientific name: *Cereus repandus*
Family: Cactaceae
Synonyms: *Cereus peruvianus*; Giant Club Cactus, Hedge Cactus
Zone: 9a

Overview:

The Apple Cactus is a columnar, tree cactus growing up to 10m (33ft) tall with a 5m (16ft) spread. Cultivated plants are usually kept at about 3m (10ft) in order to be harvested more easily. The large, beautiful flowers are white, an adaptation for night pollination, since flowers are primarily pollinated by bats. Under cultivation, however, bee pollination fosters good fruit set.

The Apple Cactus is a self-infertile cactus, requiring cross-pollination with a genotypically distinct specimen in order to set fruit. The Apple Cactus has received much research as an alternative crop in the Negev Desert of Israel, and the fruits have since earned a place in the export market of that country. Spines occur in clusters along the margins of the cacti. Spineless plants are possible, but very rare. In all cases, however, there are no spines on the fruit, making them easy to harvest.

Habitat and Requirements:

The Apple Cactus is native to northern Venezuela as well as the Netherlands Antilles. There it is found growing in a tropical, semi-arid habitat that receives between 400-700mm of annual rainfall mostly during the fall and winter months, but also receives a little rain throughout the rest of the year. It often grows in well-drained rocky soil, and requires full sun. Nonetheless, like most cacti, their seedlings will desiccate in full sun, and therefore establish themselves underneath shrubs or other cover, which provide filtered sunlight.

Although the cactus is found in such a narrow geographical region, it seems highly adaptable to various climates, and can be grown successfully outside the tropics at least as far from the equator as central California. The cacti are only mildly salt-tolerant.

Uses:

Fruit and Seeds:

Plants produce sweet fruits with an inedible, smooth, bright-red or yellow-orange rind, and white flesh similar in taste and appearance to Dragon Fruit. The seeds are embedded in the edible pulp and can also be eaten.

Stems:

The stems are also edible and can be eaten raw or cooked. They are similar to *Opuntia* pads in texture, and have a similar taste.

Propagation:

Plants can be propagated asexually from cuttings or from seed. Cuttings are very easy to root in well drained soil, and can be rooted standing up, or by laying them on their side. Roots will develop wherever the section of stem touches the soil.

Seeds are very tiny, but can be germinated easily and quickly. With fresh seeds no pre-treatment is necessary, but they may be soaked in lukewarm water overnight. Flats should be filled with a rich, humus-like potting mix that has been brought to field capacity. The seeds are surface-sown by sprinkling them on top of the tamped potting mix, and then misted, before the flat is covered with thin, plastic film. Then the trays are placed in bright, diffused light.

Most seeds will germinate within one or two weeks. The plastic may be left on the flat for up to a week afterwards and then removed. The seedlings are very delicate for about the first month and need to be misted on a daily basis. Seedlings grow well under fluorescent grow

Figure 34: *C. repandus*. Sde Nitzan, Israel
Credit: Etan J. Tal

bulbs. Under optimal conditions, seedlings will attain a height of up to 90cm (three feet) in two years.

Potential Drawbacks:
The fruits have a tendency to burst upon ripening when the plants receive water.

Nutrition:
Glucose and fructose levels in the ripe fruit pulp are about 5.5g each. The vitamin and mineral content is relatively unknown.

Related Species:
Cereus jamacaru - Native to the Bahia region of Northeast Brazil.
C. fernambucensis - Grows indigenously along the beaches of Northeast Brazil.
C. hildmannianus ssp. uruguayanus - Native to semi-arid regions of Uruguay and southern Brazil.

Kalahari Cucumber

Scientific name: *Coccinia sessilifolia*
Family: Cucurbitaceae
Synonyms: Sessile-Leaved Cucumber, Borobohlolo, Borobohlolo Melon
Zone: 9b

Overview:
The Kalahari Cucumber is a dioecious, perennial, caudiciform succulent with a caudex (modified underground bulbous stem) that can grow up to 25kg under ideal conditions. These large bulbous growths store water and are an adaptation for a hostile desert where sandy conditions prevail. The above ground growth looks very similar to a typical garden cucumber. A male plant is needed in order for female plants to set fruit. The fruits turn reddish orange and ripen during the summer months.

Habitat and Requirements:

Figure 35: *C. sessilifolia* fruit. Pilanesberg National Park, South Africa
Credit: Joan Young

Coccinia sessilifolia is native to the southwestern portion of the Kalahari Desert in the modern-day countries of Namibia, Botswana, and South Africa. This is the driest region of the Kalahari and receives between 100-400mm of rainfall per year. Some years, however, experience less than 50mm of rainfall. The Kalahari is a predominantly summer rainfall region, which receives erratic rains. Furthermore, the Kalahari is renowned for its "Kalahari Sands," which are deep sandy soils. Such conditions give rise to a plethora of caudiciform succulents, as well as plants with very deep roots.

The Kalahari Cucumber plants prefer well-drained, sandy soil and full sun, and are well-adapted to areas with extreme heat as long as there are some summer rains. In very hot and dry areas the plants can be grown in slightly shady conditions.

Uses:

Fruit and Seeds:

The fruits are mildly sweet when ripe, but are generally eaten while they are still green. Ripe or unripe fruits can be eaten raw or cooked. The unripe fruits taste similar to asparagus shoots.

Root:

The swollen root, technically a stem known as a caudex, is edible raw and cooked. Caudex stems can be very tough and fibrous, so indigenous people roast them before consumption.

Propagation:

Kalahari Cucumber is propagated by seed. The seeds should be soaked overnight in a solution of Gibberellic Acid, and then sown into seed flats. Seeds will germinate in less than one month.

Potential Drawbacks:

The plants are prone to rotting in heavy soil, especially where the soil may be cold when wet.

Figure 36: *C. sessilifolia* flower
Credit: Benny Møller Jensen

Nutrition:

Fruits	(per 100g):
Water	82.3g
Energy	61kcal
Protein	2.1g
Fat	0.2g
Carbohydrates	13g
Fiber	1.3g
Calcium	38mg
Magnesium	2mg
Phosphorous	24mg
Iron	0.2mg
Thiamin	0.19mg
Riboflavin	0.13mg
Ascorbic Acid	25mg

Source: PROTA 2: Vegetables/Légumes

Caudex	(per 100g):
Water	84g
Energy	47kcal
Protein	1.0g
Fat	0.1g
Carbohydrates	11g
Fiber	2.7g
Calcium	351mg
Magnesium	60mg
Phosphorous	40mg
Iron	2.2mg
Thiamin	0.02mg
Riboflavin	0.01mg
Niacin	0.28mg
Ascorbic Acid	6.7mg

Related Species:

Coccinia grandis is a more tropical species less tolerant of drought, which also produces edible fruits. *Coccinia sessilifolia* and *C. grandis* can be interbred.

Sanky

Scientific name: *Corryocactus brevistylus*
Family: Cactaceae
Synonyms: Guacalla, Quisco de Flores Amarillas, Sancayo
Zone: 8b

Overview:

Sanky is a columnar cactus that can attain heights of up to 7m (23ft) with a spread of up to 5m (16ft) wide. Sanky produces sharp spines that can grow up to 24cm long, but are usually only between 5-15cm in length. Yellow, tubular flowers give rise to large, light green fruits. The plant is valued for its edible fruit.

Although the fruits are covered in sharp spines, as the fruits ripen these spines fall off or are removed without difficulty. The plants are self-incompatible, and require cross-pollination with a genetically distinct individual in order to set fruit.

Habitat and Requirements:

Sanky is often found growing on rocky soils and calcareous sites on hills high up in the Andes Mountains. The plant is native to the Andes of southern Peru and northern Chile, and is found growing up to an altitude of at least 3500m (11,500ft). The plant grows in full sun.

The native habitat of Sanky is subject to rain during the summer months of December through March. The rain is rather unreliable, however, and the cacti usually only receive between 100-400mm of rainfall per year. Due to the altitude, the region does not experience extreme heat.

Uses:

Fruit and Seeds:

The fruits are sour, tasting similar to lemons, and can be cultivated wherever lemons are grown, yet are a lot more drought-tolerant, and even more cold-tolerant. Sanky fruits can be eaten plain, but are often mixed with sweeteners such as honey and used to make a refreshing, thirst-quenching beverage. The fruits are often as large as a softball. The shells of the fruit are also used as shampoo or as an ingredient in shampoo, and help stop dandruff, and are said to help prevent hair loss.

Miscellaneous uses:

Cuttings can be planted close together in order to easily form a living fence through which no humans or larger animals would dare pass.

Propagation:

Stem cuttings can be easily rooted. They should be cut using very thick gloves and a piece of carpet or burlap to wrap the section of stem being cut. A small handsaw is best used to the make the cut. Once severed, the stem cutting should be left to callous for a couple

Figure 37: *C. brevistylus*. Huanca, Peru
Credit: Carlos Zuñiga

days in cool shade. Then it can be rooted easily in a well-draining soil mix.

Seeds should be surface sown in seed trays and covered with plastic wrap for up to two weeks, as most seeds will germinate within that two week time frame. Older seeds are more difficult to germinate and may need to be soaked in warm water overnight. Direct sunlight is detrimental to seedlings during the first month or two of growth, so seedlings should be propagated in partial shade or under fluorescent lights for the first few months of growth.

Potential Drawbacks:

The spines can be excessively long, making harvesting a difficult and painful task. Fruits should be harvested using a large pair of forceps such as those used in child birth.

Nutrition:

Sanky has been scientifically proven to have antioxidant properties. The plants are widely marketed and espoused throughout Peru with the claims that the fruits have ten times more Vitamin C than an orange, twice as much potassium as a banana, and high levels of calcium and phosphorous. However, scientific evidence to back up these claims seems to be lacking. Nonetheless, the fruits are consumed widely without ill-effects.

Rounded Noon-Flower

Scientific name: *Disphyma crassifolium subsp. clavellatum*
Family: Aizoaceae
Synonym: Round-Leaf Pigface
Zone: 9b

Figure 38: *D. crassifolium subsp. clavellatum*
Santa Barbara, California

Overview:

The Rounded Noon-Flower is a spreading, succulent groundcover with round leaves and bright purple flowers. The leaves have a cylindrical shape with a pointed tip and are green during times of available moisture. During periods of drought the leaves often turn reddish until moisture returns.

Habitat and Requirements:

The Rounded Noon-Flower is native to the arid regions of Western Australia and the semi-arid Mediterranean regions of southwestern Australia to Victoria and into parts of New Zealand. The Rounded Noon-Flower grows in areas that receive between 200-800mm of rainfall per year. It is tolerant of coastal conditions as well as inland conditions and prefers full sun.

The plant is found growing in a wide array of soils including alluvial clays, saline depressions, coastal limestone cliffs, and sandy areas. The plants are highly tolerant of water-logging, and extremely tolerant of salinity. They have been measured to tolerate salinity levels of up to 16 deciSiemens per meter (dS/m).

Uses:

Leaves:

The leaves can be eaten raw or cooked, but are generally best cooked unless very young and tender.

Miscellaneous uses:

This plant is virtually fire-proof, and can be used as a buffer to prevent the spread of fires.

Propagation:

Plants are very easy to propagate. As they spread, they root along the stems at the leaf nodes, which means cuttings can be propagated without difficulty. Seeds, on the other hand, are tiny, and need to be surface sown. The seeds should be sprinkled on the surface of a moist potting mix and then misted. After misting, the seed tray should be covered with plastic wrap in order to raise relative humidity levels to near 100%. The seeds have great germination rates, and germinate within the first two weeks.

Potential Drawbacks:

Many succulents contain oxalic acid. It is unclear whether or not this species contains any harmful compounds. In any case, caution should be taken when ingesting the leaves of this plant raw, or when ingesting large quantities.

Figure 39: *D. crassifolium subsp. clavellatum* flower
Cheetham Wetlands, Victoria, Australia. Credit: Melburnian

Related Species:

In inland areas, the Rounded Noon-Flower often grows where Common Ice Plant (*Mesembryanthemum crystallinum*) grows.

Ruby Salt Bush

Scientific name: *Enchylaena tomentosa*
Family: Chenopodiaceae
Synonyms: Barrier Salt Bush, Kalaya, Emu Bush
Zone: 9a

Overview:

The Ruby Salt Bush is a small, grayish-green shrub growing up to 1m (3.2ft) tall. As the shrub ages it becomes somewhat woody, but definitely cannot be described as a woody shrub. The foliage is thin and individual leaves are cylindrical and narrow. The foliage is covered

Figure 40: *E. tomentosa*. Lake Mungo National Park, New South Wales
Credit: M. Cosgrove

with short, soft, white hairs, which give the plant its grey appearance. The flowers are inconspicuous. The plant produces its berries year round. The fruits ripen from green, yellow and orange to a brilliant, bright red and are 5-8mm in diameter.

Habitat and Requirements:

Ruby Salt Bush is native throughout Australia and grows in semi-arid and arid regions. It has a rainfall tolerance of about 100-1000mm per year. The plants prefer full sun, but are adapted to a wide-range of soil conditions (although they tend to prefer good drainage) and are salt-tolerant, tolerant of nutrient-poor soils and somewhat tolerant of waterlogging.

Uses:

Fruit and Seeds:

The fruits can be eaten raw or cooked, and have a sweet and slightly

salty flavor.
Leaves:
The leaves contain high levels of oxalic acid and should only be eaten in small quantities when cooked.

Propagation:
Plants can be propagated easily from cuttings and seeds. Cuttings should be planted into a very well-draining mix and kept moist, but not wet. Seeds require no special pretreatment and are sown 0.5-1cm below the surface of the soil. The seeds will germinate within one month.

Nutrition:
The fruits are high in vitamin C.

Midnight Lady

Scientific name: *Harrisia pomanensis*
Family: Cactaceae
Synonyms: Ulúa, Pitaya, Apple Cactus; *Cereus pomanensis, Eriocereus pomanensis, Cereus regelii, Harrisia bonplandii, Harrisia regelii, Eriocereus regelii, Cereus bonplandii, Eriocereus bonplandii, Eriocereus polyacanthus, Eriocereus tarijensis*
Zone: 9a

Overview:
The Midnight Lady is a sprawling/climbing cactus with thin stems that can grow up to 3 meters (10 feet) in length and have 4-7 ribs. The stems are 4-8cm thick and are armed with small spines. The plants produce very showy, bright white, night-blooming flowers that give rise to bright red, spineless fruits. These fruits can grow up to 5-6cm in diameter and contain a deliciously sweet, white pulp that contains many small, black seeds. Midnight Lady cacti are self-fertile and do not require cross pollination in order to set fruit.

Habitat and Requirements:

Figure 41: *H. pomanensis*. Gran Chaco, Argentina
Credit: Natalia Galasso

Midnight Lady cacti are native to the Gran Chaco, a semi-arid, subtropical region that encompasses parts of modern day Bolivia, Paraguay and northern Argentina. They grow in a region that receives between 350-650mm of average annual precipitation. Most rain falls in the summer, when temperatures commonly exceed 38°C (100°F). The climate is comparable to the Sonoran Desert of North America.

Midnight Lady cacti prefer well-drained soil and often grow in rocky areas. The cacti are somewhat shade-tolerant, and are often found growing in the understory of, and climbing light-canopied trees.

Uses:

Fruit and Seeds:

Fruits are usually dissected and their pulp is eaten. The pulp contains many small black seeds, which are also edible. The pulp tastes sweet, and is rather thirst-quenching.

Miscellaneous uses:

The sprawling habit of this cactus is useful in creating barriers to keep humans and larger animals from walking through a given area.

Propagation:

The seeds are small and can be propagated like most cacti species. The seeds can be surface sown or slightly covered with a thin layer of fine sand. The soil medium is then brought to field capacity and covered with plastic wrap. Most seeds will germinate within one month. The plants can also be grown easily from stem cuttings, and, due to their small diameter, are easy to separate from the parent plant.

Related Species:

Harrisia martinii and *H. tortuosa* both have bright red, edible fruit, and look very similar to *H. pomanensis*. *H. pomanensis*, however, is the only one of the three that has spineless fruit.

Ice Plant

Scientific name: *Mesembryanthemum crystallinum*
Family: Aizoaceae
Synonyms: Common Ice Plant, Crystalline Iceplant, Diamond Ficoides
Zone: 9b

Overview:

Ice Plant is a prostrate, spreading, groundcover succulent with thick leaves. The plant is named Ice Plant due to the large, water-holding bladder cells that grow on the outside of the leaves and stems, which glisten in the sun as if they were covered with frost. When there is enough water available, the leaves of certain specimens can grow up to 15cm in diameter and form a nice ornamental mat over the ground, which turns a reddish color during dry periods. Ice Plant produces brilliant, bright white flowers in the spring – early summer. The plant has the ability to switch from C3 to CAM (Crassulacean Acid Metabolism) photosynthesis. In areas of ample moisture plants are perennial; however, they usually grow as annuals in areas with prolonged droughts.

Figure 42: *M. crystallinum* flower. Santa Barbara, California

Habitat and Requirements:

Ice Plant is native to the Cape of Good Hope located in the southwest corner of the West Cape of South Africa. This area receives between 500-1000mm of average annual rainfall. Nonetheless, the plant has succeeded in much drier regions as well. For example, Ice Plant has naturalized in parts of the Mediterranean climate region of coastal southern California and throughout areas of western and southern Australia.

Ice plant prefers to grow in well-drained soil, but also tolerates heavy soil, and often colonizes disturbed areas. The plants are well adapted to growing in areas immediate to the coast, as well as inland areas as long as temperatures do not get too cold. They are salt-tolerant, and tend to accumulate salt in their plant parts. They require full sun. Ice Plant switches from C3 photosynthesis to CAM photosynthesis during periods of prolonged drought and in areas of very salty soil.

Uses:

Figure 43: *M. crystallinum* during the winter. Santa Barbara, California

Fruit and Seeds:
The seeds are eaten as a famine food. However, it would take lots of seeds for a substantial meal.
Leaves:
The leaves are edible raw or cooked, and taste like a saltier version of spinach. Due to their thickness, the leaves can be stored for a long time before they begin to wilt. Since the plants accumulate salts, the level of salt in the leaves depends heavily on the level of salt in the soil in which the plants are grown. Boiling the leaves helps remove some of the salt prior to consumption. Cooked leaves are great in stir-fries, whereas raw leaves make a nice addition to salads. Crushed foliage can also be used as a soap substitute.

Propagation:

The seeds are small and can be soaked overnight. The seeds from the top part of the seed pod tend to germinate more quickly than seeds found at the bottom of the seed pod. In any case, the seeds can be surface sown in flats, misted, and then covered with plastic wrap and placed in bright light. Most seeds will germinate within two weeks.

It is important not to over water, and to use a light, well-drained potting mix, because young plants are prone to root rot and damping off fungus. Also, care should be taken when transplanting young plants because they rapidly develop a widely spreading root system that can be damaged easily.

Potential Drawbacks:

It is reported that plants contain oxalic acid, and perhaps psychoactive alkaloids related to those found in Kanna (*Sceletium tortuosum*). Therefore, caution should be taken when ingesting this plant. Small quantities consumed infrequently appear to have no negative effects. Cooking the plant thoroughly may also help reduce some of the potential negative effects.

Since plants accumulate salt in their tissues that is released into the soil when the plants decompose, salt-sensitive plants may have a difficult time growing nearby. Alternatively, the plants can be used as an agent to remove salt from soils that are too saline for other plants to grow in.

Nutrition:

The plant is probably a good source of Vitamin C, as it was reportedly eaten by sailors to treat scurvy.

Indian Fig

Scientific name: *Opuntia ficus-indica*
Family: Cactaceae
Synonyms: Mission Prickly Pear, Prickly Pear, Spanish Tuna, Mission Cactus, Tuberous Prickly Pear, Smooth Mountain Prickly Pear, Nopal
Zone: 9a

Overview:

The Indian Fig is a large, trunk-forming cactus with pads that multiply, and turn woody with age. The Indian Fig can reach heights of

Figure 44: *O. ficus-indica.* Liguria, Italy
Credit: Francesco Cepolina

up to 5m (16ft) with a spread of up to 3m (10ft). The stems are in the shape of oblong pads, and are usually armed with spines and/or glochids arranged in clusters. Beautiful, bright yellow to orange flowers are produced in spring and early summer, and give rise to fruits of various colors.

Habitat and Requirements:

Indian Fig is native to Mexico, however, its exact origin is unknown. The plant is adaptable to many climates, and has naturalized in many Mediterranean and subtropical regions throughout the world. The plant also grows well in hot desert climates, and succeeds in areas that receive 150-600mm of average annual precipitation. During prolonged droughts, the stems will start to shrivel. This is an indication that plants should be watered.

The plants prefer soils that have a pH between 6.0 and 7.5. They also prefer well-draining soil, and are prone to fungal diseases when they are waterlogged. As such, they dislike clay soil, and instead prefer sandy and rocky soils.

Uses:

Fruit and Seeds:

The fruits should be brushed with a fine-haired brush before handling in order to remove the glochids. Also, a pair of forceps is useful in the handling of the fruit. The fruit can then be held over an open flame or dropped in a pot of boiling water to further destroy any remaining glochids. Fruits range from sour to very sweet, and often taste similar to watermelons. They can be eaten raw, dried or cooked, and are often juiced. The seeds are also edible, and can be ground into a meal.

Flowers:

The flowers are edible.

Stems:

The stems or "pads" of the cacti are best eaten when young before they develop their spines and glochids. They are usually diced and steamed. They are rather mucilaginous, but have an excellent flavor reminiscent of green beans. The pads are also a favorite fodder of cattle.

Miscellaneous uses:

Indian Fig cacti are useful planted on slopes to prevent erosion. The cacti can also be planted as barrier hedges.

Propagation:

The plants are easy to propagate from cuttings. A pad is removed from the rest of the plant and left to callous in a dry area out of direct sunlight for the next 1-2 weeks, and is then placed in well-drained soil.

The plants can also be propagated from seed. The seeds should be soaked for 12-24 hours in water before sowing in flats. The seeds should be surface sown, and then covered with a very thin layer of fine sand. They should be covered at a depth of no more than 5mm. The seed flat should then be covered with a thin layer of translucent plastic film in order to trap humidity. Most seeds will germinate within one month.

Potential Drawbacks:

The pads and fruit are often armed with clusters of spines and glochids.

The glochids are especially irritating because they are nearly invisible to the naked eye, sharply barbed, and plentiful. Improper handling of the plant parts can result in many glochids being lodged into the handler's skin. The glochids can persist in clothing even after it has been washed.

Nutrition:

Raw fruits (per 100g):

Water	87.55g
Energy	41kcal
Protein	.73g
Total lipid (fat)	.51g
Carbohydrates	9.57g
Fiber, total dietary	3.6g

Calcium	56mg
Iron	.30mg
Magnesium	85mg
Phosphorus	24mg
Potassium	220mg
Sodium	5mg
Zinc	0.12mg

Vitamin C, total ascorbic acid	14 mg
Thiamin	0.014 mg
Riboflavin	0.060 mg
Niacin	0.460 mg
Vitamin B-6	0.060 mg
Folate, DFE	6 mcg_DFE
Vitamin B-12	0 mcg
Vitamin A, RAE	2 mcg_RAE
Vitamin A, IU	43 IU
Fatty acids, total saturated	0.067 g
Fatty acids, total monounsaturated	0.075 g
Fatty acids, total polyunsaturated	0.213 g

Source: USDA National Nutrient Database

Related Species:

There are many species of *Opuntia*. The Galapagos Islands, alone, are home to six different species. *Opuntia* also features many species native to temperate regions. Since all species of cacti produce edible fruit, the genus *Opuntia* is no exception. Other notable species besides *O. ficus-indica* are *O. subarmata*, *O. engelmannii* and *O. vulgaris*. *O.*

cochenillifera produces tasty pads, but has small fruit. *O. tomentosa* is a tree species of Opuntia that produces good fruit. *O. joconostle* is an excellent species from Mexico, which is marketed extensively in that country.

Salicornia

Scientific name: *Salicornia bigelovii*
Family: Chenopodiaceae
Synonyms: Bigelow's Pickleweed, Dwarf Saltwort, Dwarf Glasswort, Sea Asparagus
Zone: 5b

Overview:

Salicornia is a genus of halophytic, herbaceous, succulent plants that usually grow less than half a meter in height. *Salicornia bigelovii* is an annual that only grows to about .3m (1ft) in height. It has thick, upright, segmented stems that turn reddish when they are stressed. The hermaphroditic flowers are pollinated by the wind, and produce oil-rich seeds in the autumn. This species has recently become popular in coastal areas of developing nations, due to its variety of uses and economic potential.

Habitat and Requirements:

This species of Salicornia is native to the east coast of North America. In the east of the continent the plant grows from coastal Nova Scotia down along the entire east coast of the United States, and on many Caribbean islands. In the west the plant occurs from southern California down through Baja. Since the plant is dependent on ocean water, it grows in regions with virtually no precipitation, as well as in areas that receive plentiful, year-round rainfall.

The plants are extremely salt-tolerant, and can be irrigated with water that has a salinity level of at least 49 deciSiemens per meter (dS/m). Due to its high level of salt-tolerance, the plant is often irrigated with drainage water from conventional agricultural systems as well as with seawater. Salicornia plants almost always grow where

Figure 45: *S. bigelovii.* Goleta, California
Credit: Mark Viergutz

there is standing water, yet can also grow in drier areas, as long as there is ample groundwater just below the surface.

The plants respond well to effluent water, and are often planted outside of saltwater shrimp farms in order to treat the waste water of the shrimp. In coastal Eritrea, for example, elaborate engineering projects have allowed seawater to flow inland from the ocean to passively irrigate vaste expanses of Salicornia fields grown in conjuction with shrimp farms, Tilapia farms, and Mangrove forests. Ocean water can be passed through succeeding fields of plants with high salt-tolerance in order to irrigate a variety of potential crops.

Uses:

Fruit and Seeds:

The seeds ripen in early fall. Raw seeds are inedible, due to the presence of saponins. Seeds contain approximately 30% edible oil similar to safflower oil. The Salicornia seed oil has a high linoleic acid content (about 73-75%), and the oil is commonly extracted from the seeds to be used for cooking, and also for biodiesel production. A single hectare of land planted to Salicornia can yield about 225-250 gallons of biodiesel per year. The remainder of the oil-extracted seeds, known as the seed meal, can be used as a protein-rich feed for livestock. Cooking the seeds as well as changing the water in which they are cooked, can help eliminate or greatly reduce the presence of saponins.

Leaves:

The young leaves and stems are eaten cooked as well as pickled. The entire foliage can also be used as fodder for livestock, and is commonly implemented as part of goats' diets.

Propagation:

The seeds require no special pretreatment, but may be soaked in water. Since the seeds naturally germinate in areas with ample moisture, they can be sown in seed flats that have been flooded or excessively watered. Covering trays with plastic wrap will also help increase germination rates by trapping humidity.

In commercial operations, the seeds are simply direct sown with a tractor at a depth of about 1cm into prepared fields. They are then irrigated using an overhead sprinkler system, and/or flood irrigated by releasing water from canals.

Potential Drawbacks:

Animals fed the seed meal as part of their diet should also be supplemented with cholesterol in order to counteract the antigrowth factors of the seeds' saponins. Also, the foliage should be washed before being fed to animals, in order to reduce its salt content.

Nutrition:

Leaves/Stems:	(g/100g):
Moisture	88.42g
Crude Protein	1.54g
Total Lipids	0.37g
Crude Fiber	0.83g
Carbohydrates	4.48g
Ash	4.36g
Biochemical Composition	(mg/kg):
Total Chlorophyll	569.1mg
Beta-Carotene	159mg
Ascorbic Acid	58.4mg

Mineral Elements	(mg/g):
Sodium	9.98mg
Potassium	1.76mg
Magnesium	1.18mg
Calcium	0.62mg
Phosphorus	0.18mg
Iron	0.01mg
Zinc	4.05mg
Copper	0.91mg

Amino Acid Composition of Seeds	(g/kg):
Asparagine	1.16g
Glutamic Acid	1.63g
Serine	0.68g
Histidine	0.26g
Glycine	0.53g
Threonine	0.55g
Arginine	0.68g
Alanine	0.69g
Tyrosine	0.44g
Cysteine	0.03g
Valine	0.59g
Methionine	0.09g
Phenylalanine	0.55g
Isoleucine	0.47g
Leucine	0.94g
Lysine	0.73g
Proline	0.83g
Total	10.86g

Source: Nutritional Characterization and Changes in Quality of Salicornia bigelovii Torr. during Storage

Related Species:

Nearly all Salicornia species produce edible stems. Nonetheless, few species will ever receive as much attention as *Salicornia bigelovii*. The most closely related species with the most potential to receive the same treatment as *S. bigelovii* is *S. europaea*. That is because *S. europaea* seeds also yield edible oil similar to safflower oil.

Kanna

Scientific name: *Sceletium tortuosum*
Family: Aizoaceae
Synonyms: Kauwgoed, Kougoed, Canna, Channa, Tortuose Fig Marigold
Zone: 9a

Overview:

Kanna is a low-growing, spreading, perennial succulent groundcover. The plant is drought-deciduous, and "skeletonizes" its leaves when dormant during the summer months. The plant produces beautiful white flowers in the spring. The plants spread quickly, and usually live 3-5 years. Wild populations are threatened by overharvesting, and recently much land has been devoted to cultivating Kanna in order to serve a market niche exploited by the plant's anti-depressant effects.

Habitat and Requirements:

Kanna is native to the southwestern Karoo ecoregion primarily in the Western Cape, but also in the southern extreme of the Northern Cape, and the west part of the Eastern Cape of South Africa. This distribution range experiences 97-595mm of average annual rainfall. The region is dominated by a Mediterranean climate with cool, wet winters, followed by a long dry season. Kanna plants go deciduous in the summer due to the intense sun and dry conditions of their native habitat.

The plants prefer to grow in partial shade, and, in the wild, propagate themselves most freely under shrubs. Nonetheless, the plants will still grow well in full sun. The plants prefer a sandy loam soil, but will

Figure 46: *S. tortuosum.* Santa Barbara, California

tolerate nearly all soil types, including clay.

Uses:

The entire plant, but especially the aboveground portion, is chewed and/or eaten as an anxiolytic, anti-depressant, and empathogen. The plants are harvested in October (April in Northern Hemisphere), macerated, and then left in sealed bags in the sun for one week to ferment. Throughout the week that they are fermenting, the bags should be opened and the plant material stirred occasionally. The fermenting process helps breakdown the oxalic acid present in the plant tissue. Alternatively, the fresh plant material can be roasted for about an hour in hot sand over a bed of fire.

The resultant material has very low or nonexistent oxalate content, but is rich in alkaloids. Nine alkaloids have been identified in Kanna plants. The primary alkaloids are classified as Selective Serotonin Reuptake Inhibitors (SSRI), which have known anti-depressant and anxiolytic effects. A typical dose is measured as a length of stem carrying three sets of leaves, or as 0.5-2 grams of powder. Users of higher dosages report feeling euphoria, with initial stimulation followed by sedation. In therapeutic dosages, the plant's alkaloids have been shown to treat obsessive-compulsive disorders, bulimia nervosa, alcohol and drug dependence, anxiety, and major episodes of depression.

Propagation:

Kanna seeds should be soaked in water for 12-24 hours and then sown on the surface of a seed flat, or covered slightly with a thin layer of sand or potting mix. The seed flat is then misted to field capacity, covered with plastic wrap, and placed in an area out of direct sunlight. Most seeds will germinate within one month, but germination is erratic, and some seeds may take longer. In general, germination rates are low. Cuttings, on the other hand, are almost guaranteed to root, and can be planted directly.

Potential Drawbacks:

Since Kanna is regarded as an SSRI, it should not be consumed in conjunction with other SSRIs or with MAO Inhibitors.

Although usually not a problem, the plants may succumb to diseases if they are irrigated during the summer months when they would normally be dormant in the wild.

Related Species:

Of the 24 species of *Sceletium*, the following are the only ones known to have similar alkaloid profiles as *Sceletium tortuosum*: *Sceletium sp. nova*, *S. joubertii*, *S. namaquense*, *S. strictum*, *S. emarcidum*, and *S. subvelutinum*. The genus *Delosperma*, a more distant relative, primarily from the wetter parts of southeast Africa, is a genus of mostly psychoactive groundcover succulents, many of which contain powerful tryptamines, including dimethyltryptamine (DMT).

Galloping Cactus

Scientific name: *Stenocereus gummosus*
Family: Cactaceae
Synonyms: *Machaerocereus gummosus*; Pitaya Agria, Sour Pitaya
Zone: 9b

Overview:

The Galloping Cactus is a semi-erect, sprawling, columnar cactus that attains heights of up to 1.5m (5ft). In the summer and/or in response to rain, the plant produces pink and white, night-blooming flowers that only open once. When properly pollinated, the flowers give rise to large (about the size of an orange), red fruits that contain an excellent-tasting, bright red pulp. The plant is self-infertile and requires cross-pollination with another genetically distinct specimen in order to set fruit. The cactus is slow growing and can take up to two years to attain a height of 15cm, even under nursery conditions. As the plant matures, however, its growth rate increases.

Habitat and Requirements:

The Galloping Cactus is native to Baja as well as the Isla de Tiburón and a small part of eastern Sonora in Mexico. Therefore, the cactus is well adapted to areas that receive about 75-300mm of average annual rainfall. The Galloping Cactus' native range predominantly experiences summer rainfall and can extend as far south as the cape region of southern Baja. Nonetheless, the range also extends into the north of Baja (up to Ensenada) along the Pacific coast, which is a Mediterranean-climate region that receives the majority of its precipitation during the winter. The plant tolerates extreme heat and considerable droughts.

Galloping Cacti prefer well-drained soil, and are often found growing on rather rocky substrates. The cultural requirements of this cactus are similar to most other desert species of cacti.

Uses:

Fruit and Seeds:

The fruit is often regarded as the best-tasting fruit of all cacti. Despite

Figure 47: *S. gummosus*. Palm Desert, California

the common names, the fruits have a predominantly sweet flavor. The seeds are small, edible, and barely noticeable when eating the fruits.
Stems:
Although the stems of Galloping Cactus contain a poisonous sap and cannot be eaten, they can be used to help gather food, because the sap effectively stupefies fish. Indigenous people often crush the stems of the Galloping Cactus in bodies of water in order to easily grab fish that are present.
Miscellaneous uses:
Due to this species' sprawling habit, the plants tend to form thickets. These thickets can be used as a barrier against larger animals.

Propagation:

Germination rates have been shown to improve after the seeds have passed through a bird's digestive system. This suggests that the seeds may also benefit from being soaked in diluted solutions of sulfuric acid. In any case, the seeds benefit from minor scarification, such as that produced by rubbing the seeds between sandpaper.

The seeds should be surface sown in seed flats, and then covered with plastic wrap to maintain high levels of humidity. The seedlings should be pricked out of the flat and potted up when they are at least 1cm tall. They can be planted out when they are at least 10-15cm tall. The young cacti should be grown out of direct sunlight for at least the first three months of their life.

Potential Drawbacks:

The unripe fruits are spiny, and, although the spines usually abscise when the fruits ripen, they can occasionally persist. Nonetheless, the spines are not very difficult to remove, and the fruits are not armed with glochids. Therefore, they are relatively painless to handle.

One drawback of the fruits, however, is that they do not seem to have a very good shelf life. Ripe fruits tend to ferment quickly, so should either be eaten immediately, canned, or frozen.

Related Species:

Stenocereus queretaroensis from central Mexico has been widely planted as a crop plant. The fruits have a good flavor, and the stems are also said to be edible. This plant has similar climatic requirements as the Apple Cactus (*Cereus repandus*).

Part 4: Miscellaneous Edible and Medicinal Plants

The plants showcased throughout the following pages represent the abundant botanical resources found throughout various drylands of the world. Whether we're examining the amazing Marula tree, which can produce up to 4.5 tons of fruit per tree per year with a Vitamin C content four times that of oranges, and a calorie-rich seed kernel surpassing the taste of Macadamia nuts; or the rare Lucumillo, an endangered, edible fruit-bearing shrub native to a 66km stretch of coastal Chile - the plants featured within these pages indicate that drylands host a vast array of opportunities for people seeking resourceful ways to live off of the land.

Baobab Tree

Scientific name: *Adansonia digitata*
Family: Malvaceae
Synonyms: Upside-Down Tree, Monkey Bread, Cream of Tartar
Zone: 10b

Figure 48: *A. digitata.* Tarangire National Park, Tanzania
Credit: Ferdinand Reus

Overview:

This large tree from Africa grows to a maximum height of about 30 meters (98 feet), and can achieve a trunk diameter of up to 16 meters (52.5 feet). The large, sometimes hollow trunks have given rise to a plethora of intriguing historical and contemporary uses. Baobab trees are extremely long-lived - being able to live for upwards of 3000 years. Large, white flowers are pollinated by fruit bats, and give rise to fruits up to 10 inches long. Nearly all parts of the tree are useful medicinally as well as nutritionally.

Habitat and Requirements:

Baobab trees demand well-drained soil. They are particularly susceptible to rot in areas with cool, wet winters. Since they are widely dispersed throughout areas of the arid, African tropics, the trees have not evolved to withstand frost.

Although trees are unable to withstand cold, inundated soil and light frost, they are extremely well adapted to hot, drought-prone regions with sandy soils. The trees are seldom found in mountainous areas, and instead can be most often seen in wide-open savannas where there is a high ground-water table. Plentiful rain years are required for wild trees to germinate and nurse themselves into maturity. Once mature, however, Baobab trees can grow in areas that receive only 90mm of annual rainfall. The high end of their rainfall tolerance is about 1500mm of annual rainfall. In these areas trees need very well-drained soil to avert disease and fungal pathogens. The trees need full sun.

Under cultivation, trees respond well to compost and bone meal mixed with the local, well-draining soil. Supplemental irrigation during the first few years of growth can hasten the growth rate and increase the survival rate of the trees. Under ideal conditions, trees will start to bear fruit at eight years of age, yet it is not uncommon for trees to start bearing fruit until 12-15 years of age.

Uses:

Fruit and Seeds:

The fruit contains an edible pulp, high in Vitamin C, and is often added to milk or water to make a refreshing drink. The fruit pulp helps prevent diarrhea. The seeds, which are usually roasted, are high in protein and contain edible oil in the kernel.

Flowers:

Flowers are high in protein, and are a favorite of livestock.

Leaves:

Leaves are eaten dried and/or cooked as a vegetable, providing the consumer with sugars, calcium, potassium tartrate, and Vitamin C. The leaves are also a highly nutritious livestock fodder.

Wood:

The wood may be chewed to quench thirst.

Bark:
The bark is commonly harvested, without apparent detriment to the tree, for its strong fibers, which are made into an array of goods such as ropes, clothes, and fishing nets.

Root:
The roots of young seedlings can be eaten. Similarly, the recently germinated "sprouts" can be eaten.

Miscellaneous uses:
The massive branches create areas where water pools, which can be used as a source of drinking water. Occasionally, trees are intentionally hollowed out so that they can be used as massive water storage tanks. However, Baobab trees will often naturally have hollow trunks. These trunks have been used as beehives, jail cells, houses, bars, and bus stops.

Propagation:

The Germination of Baobab trees can often be erratic, taking anywhere from 2 weeks to six months. To increase germination rates, scarring may be practiced, although caution must be exercised as seeds can easily perish if cut too deep. Alternatively, the boiling method may be applied by quickly submerging seeds in boiling water for up to 30 seconds. This treatment can be followed by a 24-hour soak in ambient temperature water.

Baobab trees may also be propagated from cuttings.

Potential Drawbacks:

Young trees less than three years old are often killed by the elephants that eat them. Likewise, all young trees should be adequately protected from livestock and wild animals.

Nutrition:

The fruit is reported to contain large amounts of anti-oxidants, has about 50% more calcium than spinach, and three times the vitamin C as an orange. The leaves are also rich in vitamin C, and contain potassium as well as calcium.

Related Species:

Adansonia gregorii, the only species of Baobab native to Australia, has undergone trials in that country to assess its potential as a new alternative vegetable crop in Western Australia. *A. gregorii* is used in all the same ways as *A. digitata*.

Argan Tree

Scientific name: *Argania spinosa*
Family: Sapotaceae
Synonym: Moroccan Iron Wood
Zone: 9b

Figure 49: Goats browsing *A. spinosa*. Taroudant, Morocco
Credit: Katharine L'Heureux

Overview:

This keystone species of southwestern Moroccan ecosystems is a long-lived evergreen tree (200-300 years), which reaches a maximum height of 10m (33ft) with a nearly equal spread. The plants are drought-deciduous, losing all of their leaves for years at a time during the most extreme droughts. Small greenish-brown fruits up to 5cm in length are renowned for the high-quality oil yielded by their almond-sized seeds.

Habitat and Requirements:

The Argan tree is native to southwest Morocco, where it receives an average annual rainfall between 100-500mm during the winter months. Its native habitat receives no rain during the summer. The Plants are extremely drought-tolerant, and can survive prolonged droughts, which last many years by shedding their foliage and remaining in a state of dormancy until adequate moisture forces them back into production. This adaptation allows the trees to survive even when rains fail to come.

Argan trees are found on many well-drained soil types, except where there are areas of shifting sands. The trees are particularly well-adapted to calcareous and rocky soils, allowing them to inhabit rocky areas on hills with little fertile soil.

The trees require full sun. It is reported that trees have survived brief periods down to 19°F (-7°C) in protected locations near concrete structures, suggesting that the Argan trees' cold tolerance is underestimated.

Uses:

Fruit and Seeds:

Peak production of fruits occurs when the trees are around 60 years old. The fruits are edible, but have an unpleasant, bitter taste, so they are often fed to goats.

The seed, on the other hand, contains delicious, nutrient-dense, oxidation-resistant edible oil with important medicinal qualities. The fatty acid content of the oil is comparable to olive oil, yet Argan oil is more stable against oxidation. The oil is pressed from the seed and used for cooking and to treat high cholesterol, indigestion, poor circulation, and a variety of skin ailments such as stretch marks, acne and chicken pox. Due to these latter uses, the oil has promise in the cosmetics industry, and many companies are exploiting its use in anti-wrinkle creams and other products, which is likely to outstrip the supply of oil for native Moroccan people. Indeed, Argan oil is worth about 10 times more than olive oil on the world market.

The leftover cake from the seeds after pressing the oil is used as a staple food.

Leaves:
The leaves as well as the fleshy part of the fruit can be used as livestock fodder. Goats are notorious for precariously climbing to the very tops of mature trees, voraciously pursuing a meal.

Wood:
The wood of Argan trees is very useful, and has contributed to the decline of the natural forests. The wood is useful as fuel and as timber, since it is dense and insect resistant. It can be made into charcoal or burned as firewood, or it can be used to create utensils and to build structures. The trees can be coppiced, but are very slow-growing.

Miscellaneous uses:
Argan trees are useful in apiculture, contributing to high-quality honey production.

Propagation:

Although whole fruits are often fed to goats, and the seeds are then picked out of their dung, there appears to be no necessity of acid scarification when propagating Argan tree seeds. Instead, up to 95% germination rates can been achieved as a result of a 3-month long cold stratification followed by a bath of liquefied gibberellic acid hormone.

Argan trees may be grafted in order to select reliable varieties instead of waiting for the six years necessary for fruit set from seed.

Potential Drawbacks:

Young *Argania spinosa* plants have been known to die from fungal infections attributed to the species *Fusarium oxysporum*. This is likely to be the result of overly humid conditions, stagnant air, and/or heavy soil.

Nutrition:

Argan oil has been found to contain moderate amounts of vitamin A, and high amounts of vitamin E. It also contains 80% polyunsaturated fatty acids.

The major fatty acids include:

Name of Fatty Acid:	Percent Fatty Acid in Oil:
Myristic acid	4.3%
Palmitic acid	13.5-13.9%
Linolenic acid	4.6%
Stearic acid	5.6-5.7%
Oleic acid	45.2-46.9%
Linoleic acid	31.5%

Source: *Minor Oil Crops*

Purple Hook Berry

Scientific name: *Artabotrys brachypetalus*
Family: Annonaceae
Synonym: Large Hook Berry
Zone: 10a

Overview:

This evergreen plant has a variety of different growth forms. It can grow as a shrub, a small tree, or as a climbing, woody liana. The maximum height of the liana can be up to 10m (33ft) whereas the trees and bushes range in height from 1-5m (3-16ft). Plants have thick, leathery leaves and yellowish flowers, which give rise to fleshy, red berries that turn dark purple when ripe.

Habitat and Requirements:

Purple Hook Berry plants are native from the Transvaal in South Africa, through Botswana and Zimbabwe, and up north into Tanzania. Throughout this range, the species experiences a mean annual rainfall between 300-1200mm. Plants tolerate partial and full sun and are usually found in riparian corridors and arid woodlands. The plants prefer well-drained, sandy or rocky soil, but are also found growing on alluvial soils.

Uses:

Fruit and Seeds:

Figure 50: *A. brachypetalus*. Sani Beach Lodge, Malawi
Credit: Günter Baumann. East African Plant Database

The fruits are fleshy and sweet, and can be eaten raw or cooked.
Bark:
At least the stem bark and roots contain acetogenins, very effective anti-fungal agents. Acetogenins have also been proven to have anti-cancer effects.

Propagation:
Fresh seed should be used for propagation. Sow during the spring into a rich potting mix. Before sowing, the seeds should be soaked in water for 12-24 hours in order to imbibe moisture and hasten germination.

Potential Drawbacks:
At least one acetogenin, annonacin, has been linked to Parkinson's disease-like symptoms in rats. Consuming fruits from the Annonaceae family in large quantities on a regular basis might also contribute to similar symptoms in humans.

Related Species:

Artabotrys monteiroae is very closely related, and is used in the same ways as *A. brachypetalus.*

Desert Date

Scientific name: *Balanites aegyptiaca*
Family: Zygophyllaceae
Synonyms: Soapberry Tree, Thorn Tree, Simple Thorned Torchwood, Lalob Tree
Zone: 10a

Overview:

This hardy, multi-purpose tree is unrelated to traditional dates (*Phoenix dactylifera*). The Desert Date is a spiny, semi-deciduous (in most cases they are evergreen) tree reaching up to 10m (33ft) in height. The Desert Date has a long taproot several meters in length. The plant yields a plethora of useful parts, and is a very important economic tree valued by the people throughout its native range.

From seed, the Desert Date begins fruiting after five to eight years, and enters optimal production around 20-25 years of age. The tree continues to fruit even during periods of drought, which has contributed to their protection by native peoples who recognize their economic importance. They are long-lived trees capable of surviving for more than 100 years even in extremely hostile climates.

Habitat and Requirements:

The Desert Date is native to northern Africa, in and around the Sahara Desert, from Mauritania to Somalia. It also occurs (perhaps introduced) throughout arid and semi-arid regions of Africa as far south as Zimbabwe. It has been introduced into Arabia and India as well as certain Caribbean islands such as Curacao where it has become invasive at the dry eastern end of the island. It thrives in areas with very intense, unfiltered sunlight as well as high heat.

Throughout its distribution range, the plant grows where average annual precipitation ranges from 100-1000mm. Most areas where the Desert Date occurs, however, receive between 200-800mm average annual precipitation. Plants in areas with the minimum rainfall requirement of 100mm require a high water table and are often found on the edges of seasonal ponds or other areas prone to occasional inundation. Most of its native range experiences sporadic rain during the summer months, with very dry winters. Nonetheless, certain areas of its range such as Mauritania can experience winter rain.

The Desert Date is well-adapted to many soil types including heavy clay soil and sand. Very deep, shifting sands, however, don't seem to support this species.

Given the fact that Desert Dates grow up to 1,800m (5,905ft) in the Ahaggar mountains of southern Algeria where temperatures drop below freezing, Desert Date should be able to tolerate at least light frosts.

Figure 51: *B. aegyptiaca*. Ein-Gedi, Israel
Credit: Amram Eshel

Uses:

Caution: Plant parts contain saponins and sapogenins. Potential health impairments could occur as a result of improper or excessive consumption.

Fruit and Seeds:

Fruits are highly variable in taste, and can be sweet. Generally, however, they are more bitter and less sweet with a lower sugar content. This tree probably won't replace the sweet fruit-producing plants commonly esteemed throughout the world, but it has the advantage of being able to grow in extreme environments where these other fruiting trees cannot grow. Desert Dates can be eaten raw or dried. Desert Dates are also enjoyed by livestock and other animals.

The seeds are high in protein, oil and carbohydrates. They must be cracked and then processed to remove their bitter constituents. Generally, they are processed by boiling them in two separate changes of water, or they are steeped for 2-4 days. The oil can be extracted from the seeds and used as a high-quality cooking oil with a high stability against oxidation, while the seedcake is made into a protein-rich food than can be eaten by humans and livestock. The oil can also be used in the manufacture of soap, and the entire fruit as well as other plant parts including bark and leaves can be used as a soap-substitute by pounding and agitating in water.

Flowers:

The flowers are boiled and eaten.

Leaves:

The leaves are eaten cooked and have a high protein content.

Resin:

The resin/sap of the tree is often chewed on as a snack.

Wood:

The wood is very dense and insect resistant. Therefore, it is usually made into tools. It is one of the best fuelwoods for indoor use because it burns hot and produces little smoke. The wood can be used to make charcoal, and the shells of the seeds can also be used as a fuel source.

Bark:

The bark can be eaten in times of famine, and is also a source of strong fiber.

Miscellaneous uses:

Small twigs are used to clean teeth. Trees are spiny and can be coppiced and pollarded. Often coppicing increases the occurrence of root suckers. Such a growth habit contributes greatly to the Desert Date's use as a living fence for livestock. While consumption of Desert Dates appears to have no obvious negative side-effects with humans, the fruits can be used as an effective snail-killer. The fruits can be steeped in water at a ratio of about 10 fruits per gallon of water in order to be used as a snail poison. The bark and fruit can also be used to kill fish.

Propagation:

Desert Dates can be propagated via seeds, cuttings or root suckers. In general, it is best to direct sow seeds, because the plants rapidly develop a deep taproot, which is usually damaged by container culture or transplanting. Seeds that have passed through an animal's digestive tract yield the best germination rates. However, other methods can be utilized to achieve germination rates of up to 60%. The seeds can be boiled for 5 to 10 minutes, and then left to soak for 8-24 hours.

When they are sown directly, care must be taken to protect the seedlings from browsing by animals. Also, they should be weeded since heavy weed growth may block out all their light. When young, the seedlings appear to grow slowly because most of their energy is being put into their deep taproots to mine water from the subsoil.

Potential Drawbacks:

The Desert Date has proven to be invasive on the island of Curacao. Caution should be taken when introducing this tree to other parts of the world. The vigorous root suckers produced by this species might turn out to be a maintenance nightmare in areas where the trees are not desired.

The Desert Date was once sought after for its sapogenin, diosgenin, as a natural chemical precursor to many of the steroidal drugs on the market today. Synthetic production of these steroids destroyed the potential market for that aspect of the trees. The saponins and sapogenins contribute to the Desert Date's bitter qualities and may even have adverse health risks associated with their consumption. More research needs to be carried out regarding these risks. In general, however, most literature does not mention any adverse health effects attributed to indigenous people who eat the Desert Date. Similarly, studies of rats being fed the Desert Date seed meal revealed no major toxic symptoms in the rats.

Nutrition:

The amino acid content of the seed is almost as nutritious as that of the peanut.

Seed Kernel:	
Water	2.3-4.5%
Protein	26-30%
Oil	46-51%
Carbohydrates	15.4-17.9%
Fiber	2.4-3.3%
Ash	2.8-3.3%
Calcium (mg/100g)	102-108mg
Iron (mg/100g)	6-7mg

Seed Oil Fatty Acids:	
Total Unsaturated	63.2%
Total Saturated	36.8%
Oleic	30-40%
Linoleic	40-48%
Stearic	9-10%
Palmitic	10-12%

Source: *Non-timber Uses of Selected Arid Zone Trees and Shrubs in Africa*

Related Species:

Balanites roxburghii and *B. rotundifolia* have similar properties to *B. aegyptiaca*.

Velvet Sweetberry

Scientific name: *Bridelia mollis*
Family: Phyllanthaceae
Synonyms: *Bridelia scandens*; Velvet-Leaved Bridelia, Velvet Strawberry
Zone: 10a

Overview:

The Velvet Sweetberry is a small tree/large shrub with beautiful leaves similar in appearance to Cherimoya leaves. The leaves, however, are generally smaller than Cherimoya leaves and have a velvety appearance and texture. The trees are monoecious, and produce inconspicuous yellow-green flowers between the months of November and February (May – August in the northern hemisphere), which give rise to green berries. As the berries ripen, they turn black and become softer and sweeter.

Habitat and Requirements:

The Velvet Sweetberry is native to southern Africa from the Waterberg Mountains in Northern South Africa to southern Zambia. It has a west-east range from Namibia to Mozambique. The plant grows in areas with winter dry seasons, and summer rainfalls that range from 200-1000mm.

The plants grow in full sun, and require well-drained soil. In the wild they often grow on rocky or sandy hills and valleys where the soils are shallow.

Uses:

Fruit and Seeds:

The fruits are fleshy and sweet when they are black.

Propagation:

The Velvet Sweetberry seeds require no special pretreatments to germinate. Seeds can be placed into seed flats at a depth of 2-3cm and transplanted to nursery pots when they develop their third true leaf, taking care not to damage their roots. A well-drained (up to 50%

Figure 52: *B. mollis*. Matobo National Park, Zimbabwe
Credit: Bart Wursten

perlite or sand) but rich potting medium should be used at all stages in the nursery. Containers should be at least 12cm in diameter, because Velvet Sweetberry has a somewhat spreading root system that has evolved to bind loose soils. Narrow containers (less than 8cm in diameter) encumber the root growth of Velvet Sweetberry and are not ideal for propagation.

Potential Drawbacks:

More breeding work needs to be done to select reliable cultivars, analyze the nutrition content of the fruit and to introduce this plant into cultivation. Even very little information exists regarding this species' performance in the wild. Plants should not be planted in heavy, clay soil.

Num Berry

Scientific name: *Carissa bispinosa*
Family: Apocynaceae
Synonym: Num-Num
Zone: 9b

Figure 53: *C. bispinosa*. Lyon, France. Credit: Daderot

Overview:

The Num Berry is a relative of the more widely known Natal Plum (*Carissa macrocarpa*). The Num-Num is a small, sprawling, spiny shrub, which occasionally grows as a tree up to 4 meters (13 feet) tall. It has thick, dark green leaves and bright white flowers. These white flowers give rise to small green fruits, which ripen in late summer/early fall to a dark, crimson-red color and are notably sweet. The shrub flowers from October – March (April – September in the Northern Hemisphere), and fruits from March – October (September – April in the Northern Hemisphere).

Habitat and Requirements:

The Num Berry is native to South Africa from the West Cape through the East Cape. The shrub is mostly found in coastal areas, however, it is widely dispersed throughout southern Africa, and is found growing naturally from Mozambique in the east to Namibia in the west. Num Berry plants are very adaptable and have an annual rainfall preference between 200-1000mm.

The plant occurs in forests as well as scrub habitats, and is mildly salt-tolerant allowing it to thrive adjacent to the ocean in the Mediterranean climate region of the southwestern part of the West Cape. Plants tolerate full sun as well as partial shade, and prefer well-drained, often rocky soil.

Uses:

Fruit and Seeds:

The ripe fruits are delicious and are eaten raw or cooked or made into preserves. The entire plant, including the unripe fruits, contains bitter, milky latex, so fruits are best consumed when they are very ripe.

Roots:

Among indigenous people, the roots are used to treat toothaches.

Miscellaneous uses:

Since the plants are spiny, evergreen and usually grow to about head height, they make an excellent barrier and privacy hedge.

Propagation:

Num berry plants may be propagated by cuttings, but are extremely difficult to root. More commonly, plants are propagated by air-layering, ground-layering, shield budding, or are simply propagated from seed. Fresh seed is very easy to propagate. Using a rich potting mixture, seeds are sown just below the surface of the soil. The majority of seeds should germinate within a couple of weeks.

Potential Drawbacks:

Some plants are self-infertile and need to be cross-pollinated with other plants in order to produce fruit. Also, the fruits do not store very well.

Figure 54: Fruits of the closely related *C. macrocarpa*.
Santa Barbara, California

Nutrition:

Although smaller, the nutrition content of *Carissa bispinosa* fruit is probably similar to that of *Carissa macrocarpa*.
C. macrocarpa fruits (per 100g):

Water	84.17g
Energy	62kcal
Protein	.50g
Fat	1.30g
Carbohydrates	13.63

Calcium	11mg
Iron	1.31mg
Magnesium	16mg
Phosphorous	7mg
Potassium	260mg
Sodium	3mg

Vitamin C, total ascorbic acid	38mg
Thiamin	.04mg
Riboflavin	.06mg
Niacin	.2mg
Vitamin A, IU	40IU

Source: USDA National Nutrient Database

Related Species:

Carissa bispinosa is closely related to *Carissa macrocarpa* (Natal Plum) and *Carissa haematocarpa*. *C. macrocarpa* is more widely know and has received more research attention as an alternative crop, but is from parts of South Africa, that receive more annual rainfall. *C. haematocarpa* is slightly more drought and cold-tolerant than *C. bispinosa* and is well-adapted to inland locations. The best potential yield could be achieved by grafting *C. macrocarpa* scions onto *C. bispinosa* or *C. haematocarpa* rootstock.

Khat

Scientific name: *Catha edulis*
Family: Celastraceae
Synonyms: Bushman's Tea, Gat
Zone: 9b

Overview:

Khat is a deciduous, large shrub to small tree growing up to 10m (33ft) tall with up to an 8m (26ft) spread. The plants usually grow much smaller than these dimensions and are often maintained as a hedge at a certain height under cultivation. In areas of drought, the trees are often found growing as shrubs, whereas in areas with ample moisture, Khat plants reach their maximum height. The leaves have serrated margins and turn a beautiful yellow color in the autumn. The plants are well-known for their stimulant properties, and have been the subject of much legal controversy.

Habitat and Requirements:

Khat most likely originated in Ethiopia, but has naturalized throughout many parts of East Africa and into the Arabian Peninsula, where it is especially popular in Yemen. Khat grows throughout East Africa from Egypt down through the East Cape and into the West Cape of South Africa. Since Khat grows throughout such a wide range, it experiences many different climates and rainfall patterns. It grows in summer rainfall regions, but also in winter rainfall regions, and experiences an average annual rainfall range of 400-1200mm or more.

Khat prefers to grow in well-drained, rocky or loam soil in full sun. Plants tolerate extreme heat, and are only mildly frost-tolerant.

Uses:

Leaves:

The leaves are commonly chewed as a stimulant as they contain the alkaloids cathine and cathinone, which impart onto the user a mild feeling of euphoria, alertness, and clarity of thought. Cathinone is the primary active alkaloid of the two, and seems to be lost as the leaves dry out. Therefore, fresh leaves are preferred in order to achieve the desired effects.

Wood:

The wood is fine-grained and hard, so that it is well-suited for furniture, fence posts, and firewood.

Bark:

The bark is used as an insect repellant, suggesting that the wood itself may also have termite-resistant properties.

Propagation:

Khat is usually propagated from root cuttings and seeds. Root cuttings should be placed in a well-drained soil mix and kept on the dry side until vegetative, above-ground growth appears. Fresh seeds are best used, and require no special pretreatment. They can be directly planted or germinated in seed trays with bottom heat. The seeds should be buried only slightly.

Potential Drawbacks:

Cathinone has been listed as an illicit substance in many countries. Unfortunately, that also means that the act of growing the plant itself

Figure 55: *C. edulis*. Kenya. Credit: Spencer Woodard

may be vilified in certain areas. Cathinone has a side-effects profile similar to amphetamine, and should therefore be used responsibly. In general, more frequent and heavy use of Khat preparations results in more pronounced side-effects.

Nutrition:

Aside from the alkaloids, cathine and cathinone, Khat is also said to contain high levels of ascorbic acid (Vitamin C).

Tick Berry

Scientific name: *Chrysanthemoides monilifera subsp. monilifera*
Family: Asteraceae
Synonyms: Boneseed; Bietou, Bitou Bush (*Chrysanthemoides monilifera subsp. rotundata*)
Zone: 9b

Overview:

The Tick Berry is a fast-growing, perennial, drought-tolerant shrub with succulent-like leaves. It is a pioneer plant, and is often one of the first species to colonize areas after disturbances such as fires. It is a beautiful relative of the daisy, and has bright yellow flowers blooming throughout late fall/early winter. The flowers give rise to small, fleshy, sweet, edible berries that turn from green to brown to blackish-purple when ripe in late winter. The shrubs tend to lose productivity, and become rather scraggly after 8-12 years. The plants grow up to 2m (6.5ft) tall and can spread laterally up to 4m (13ft).

Habitat and Requirements:

The Tick Berry is native to the Western Cape of South Africa, and grows in areas that receive between 300-1000mm of average annual rainfall. They are only slightly frost-tolerant and are most commonly found growing in coastal regions with Mediterranean climates where rains are concentrated in the winter months and may be absent entirely during the summer. As a predominantly coastal shrub, the plants should be considered intolerant of extreme temperatures.

The plants prefer full sun, but can also grow in areas that receive partial shade, and are sometimes found growing in the understory of light-canopied forests. Tick Berry shrubs prefer well-drained soil, and are often found growing in rocky and sandy areas.

Figure 56: *C. monilifera subsp. monilifera*. San Marcos, California

Uses:

Fruit and Seeds:

The shrubs produce edible, fleshy berries that can be eaten raw.

Leaves:

The indigenous people of southern Africa used an infusion of the leaves as an enema to treat fevers.

Propagation:

Propagation is best achieved by seed. In the wild, birds eat the fruits whole, subjecting the hard seed coat to the acid of their digestive system. Propagation is rather difficult, since the seed coats tend to be hard and thick. Suitable pretreatments include acid scarification and mechanical scarification before soaking seeds in water overnight.

Fresh seeds should be sown in flats filled with well-drained potting mix in late winter/early spring. Seedlings should be pricked out when they develop their third true leaf, and are best left in a larger container to develop throughout the summer before being planted in the ground at the start of the next rainy season. Seeds appear to lose at least some of their viability after two years.

Potential Drawbacks:

The Tick Berry is an invasive species in Australian and New Zealand, and has become a serious ecological problem. Caution should be undertaken when introducing this species (or any of the species in this book) to new areas.

Related Species:

Chrysanthemoides monilifera subsp. rotundata is very similar, but less suitable for cultivation. *Chrysanthemoides incana* has grayish leaves and is best planted adjacent to beaches or other areas along the coast.

Kei Apple

Scientific name: *Dovyalis caffra*
Family: Salicaceae
Synonyms: *Aberia caffra*; Umkokolo, Umkolo
Zone: 9a

Overview:

The Kei Apple is a large, dioecious, evergreen shrub growing up to 9m (30ft) tall with a 7.5m (24.5ft) spread. The plants grow at a rate of about 600mm (2 feet) per year. For production the plants should be spaced at about 4m (13ft). As a hedge the plants can be spaced at 1m (3.2ft). The Kei Apple produces yellow flowers in the spring, which give rise to bright yellow fruits up to 4cm in diameter. The fruits, known as Kei Apples, are deliciously sweet with a tangy, acid flavor. The tree is a heavy producer, and is rather ornamental when

Figure 57: *D. caffra*. Santa Barbara, California. Credit: Mark Viergutz

the branches are loaded with fruit between February and April (August – October in the Northern Hemisphere). The plants are covered with long spines growing up to 7.5cm (3in) long. Plants are allelopathic, and tend to suppress weed growth by chemically stunting nearby plants.

Habitat and Requirements:

The Kei Apple is native to the East Cape of South Africa around the Great Kei River, but has naturalized throughout much of South Africa, and north through eastern Africa. The plant has also been introduced and cultivated throughout many regions of the world. Although the plant's native habitat receives between 600-1000mm of average annual rainfall, Kei Apples are found growing in areas where there is much less rain. They are particularly successful in Mediterranean climates, and probably produce well even in areas with as little as 300mm of average annual rainfall.

The plant grows in nearly all soil types, but dislikes water-logging and does not grow in areas where there is a high water table. Kei

Apples are salt-tolerant and easily withstand sea spray and coastal conditions as long as they are planted well above sea level such as on cliffs. The plant grows well in full sun as well as light shade, fruits best in dry summer climates, and is also tolerant of extreme heat.

The crowns of the trees should be pruned for optimal fruit production. Also, a one meter high buffer zone should be pruned from the ground up so as to allow easy access to the fallen fruits beneath the trees. Fruits have a good shelf-life, and can even stay intact for days on the ground.

Uses:

Fruit and Seeds:
The ripe fruits can be eaten raw or made into preserves. The unripe fruits can be used as pickles.

Leaves:
The leaves can be used as livestock fodder. Due to the spines however, animals will avoid browsing the trees, unless there is no other available fodder.

Miscellaneous uses:
The plant can be planted at a spacing of 1-1.5m (3.2-5ft) as a barrier hedge. Since the plants are armed with a plethora of very sharp, very long spines, no large animals dare to pass through the hedge.

Propagation:

Kei Apples can be propagated by air layers, softwood cuttings, budding, and from seed. Seed is not ideal, however, because of the variability of gender and fruit quality. Ideally, one male should be planted for every 20-30 females. Fresh seeds are easy to germinate. They may be soaked overnight in water and sown in seed flats.

Kei Apples are easily grafted, which allows cultivar selection and inferior seedling replacement to be easy practices. Also, Kei Apples can be grafted with an entirely different genus, *Flacourtia*.

Potential Drawbacks:

Plants coppice and may be difficult to eradicate in areas where they are not wanted. Furthermore, excessive pruning appears to contribute to an increased presence of spines.

Nutrition:
The fruits are high in Vitamin C and potassium, containing from 80-120mg of Vitamin C and 600mg of potassium per 100g of fruit. The sugar content is between 15-18% and pectin levels are about 4%. Protein content is less than 1%.

Related Species:
Dovyalis zeyheri – Wild Apricot
Dovyalis rhamnoides – Common Sour Berry

Cape Ebony

Scientific name: *Euclea pseudebenus*
Family: Ebenaceae
Synonyms: Black Ebony, False Ebony, Wild Ebony
Zone: 8b

Overview:
The name of this tree species is derived from its heartwood, which is pitch-black in color. This dioecious tree occasionally grows up to 9m (30ft) in height, but is usually smaller. It has dark, rough bark, slender leaves, small, cup-like, yellow flowers, and red fruits that turn black when ripe. The branches of Cape Ebony are rather droopy. The tree is mainly used for its sweet, edible fruits, and its high-quality wood, which is comparable to that of True Ebony, *Diospyros ebenum.*

Habitat and Requirements:
Cape Ebony is found growing indigenously in the Richtersveld, a transfrontier park straddling the border between the Northern Cape of South Africa and southern Namibia. The park is cut by the Orange River, which is used as the international boundary between the two countries. As the Orange River meanders through the parched landscape of the Richtersveld, which only receives between 30-150mm of rainfall (concentrated in the winter) per year, the Cape Ebony tree is found dotting its banks.

Figure 58: *E. pseudebenus*. Richtersveld National Park
Credit: Fred Triep

The plant is adapted to growing in areas with deep, alluvial soils where there is a high water table with plentiful groundwater. It is strictly a riparian tree that is commonly found along the Orange River, which flows year round through the Richtersveld.

The Richtersveld is an area that receives occasional fog drip from the Atlantic Ocean during the summer. These occasional fog banks give way, however, to extremely arid and hot conditions. Indeed, the Richtersveld became the home of the hottest record temperature of the entire country of South Africa when temperatures in the shade reached 53°C (127.4°F). Nonetheless, the area also experiences cold temperatures during the winter. During winter nights temperatures may drop below freezing.

Uses:

Fruit and Seeds:

When fruits are ripe they are a dark red to black color. The fruits are about 5-10mm in diameter and can be eaten raw or cooked. They are sweet with a slight astringent taste. The seeds should not be eaten.

Wood:

The wood of Cape Ebony is very similar to that of True Ebony. As such, it is harvested locally as firewood. The core of the tree is made up of a very dense, pitch black wood that can be used for carving, and is often used to make furniture, utensils, and art.

Miscellaneous uses:

Root infusions are used to treat headaches and toothaches. Also, small sections of roots and branches can be chewed on for dental hygiene as a toothbrush substitute.

Propagation:

Seeds require no pretreatment, other than soaking in water for 1-2 days prior to sowing. They can be started in seed flats in the nursery, but much caution should be taken when transplanting the seedlings into larger containers or out into the field. Seedlings 2.5cm (1in) tall often have taproots that are up to 35cm (14in) deep. The taproots do not seem to be very fragile, and can resist transplanting. Nonetheless, seeds should be direct sown in order to avoid complications due to container culture.

Related Species:

Euclea crispa is a smaller, dioecious tree from eastern South Africa. *Euclea divinorum* is dioecious and has a wider distribution range throughout southern Africa. *Euclea natalensis* is a larger, dioecious tree that grows from the West Cape, throughout eastern South Africa and far north into eastern Africa. *Euclea racemosa* is a large, dioecious shrub that grows well in coastal Mediterranean climates.

Fiori

Scientific name: *Grewia tenax*
Family: Malvaceae
Synonyms: Small-Leaved White Raisin, White Cross-Berry
Zone: 9b

Figure 59: *G. tenax*. Arandis, Namibia. Credit: E. Erb

Overview:

Fiori is a multi-stemmed, evergreen shrub growing up to 2m (6.5ft) tall with an equal spread. It has an aggressive, spreading root system. The species name, *tenax*, refers to the plant's tenacious growth habit. The smooth bark is grey and fibrous, which makes it difficult to break branches and twigs from the plant without having them fray. The plant produces alternate leaves that are nearly circle in shape. The flowers are white and give rise to small, edible fruits that turn reddish orange when ripe.

Habitat and Requirements:

Fiori is found throughout arid and semi-arid Africa and the Arabian Peninsula. The shrub grows wild in areas where the annual rainfall is between 100-1,000mm. The plant is most commonly found at the edges of deserts bordering with dry savannas, and on the rocky outcroppings of hills and mountainous slopes. Fiori is salt-tolerant, and well-adapted to growing on sandy and rocky soils due to its aggressive, spreading root system that grips loose soils tightly, combating wind and water erosion. Although the plant grows well on loose soils, it also seems to tolerate clay soil in some cases. Fiori grows best in full sun.

Uses:

Fruit and Seeds:

The fruits are sweet and can be eaten raw or cooked. They are also made into a refreshing drink, and can be fermented to make an alcoholic beverage.

Leaves:

Leaves can be eaten by livestock.

Wood:

The branches can be used for firewood and also in charcoal making. The wood is also rather flexible and can be used to make weapons such as clubs and bows.

Bark:

The bark can be boiled with water and then left to cool before it is used to get rid of lice and other insects that may live amongst the hair of humans and animals. The bark is very fibrous and can be used to make rope. Small branches and twigs can be removed from the plants, their ends frayed, and then used to clean teeth.

Miscellaneous uses:

The plant can be used as an erosion control agent in dry areas with erratic rainfall. The plants can help colonize sand dunes as well as help prevent landslides on steep slopes. The shrubs can also be planted close together and pruned as a hedge.

Propagation:

Seeds can be direct sown or planted in containers that are wide enough to accommodate their spreading root system. Seeds require no special pretreatment, but may be soaked overnight in water to help soften their seed coats. Most seeds will germinate within two months of sowing. They can be sown in flats and then pricked out when they are large enough to handle. Narrow containers might cause some level of stunting, but containers that are 8cm in diameter or wider appear to have no detrimental effect against the plants.

Nutrition:

Fiori fruits have been shown to protect the liver from diseases and toxicity. The fruit contains high amounts of iron, as well as large amounts of many amino acids, and is used to help treat anemia and malaria.

Nutrition content of fruits (per 100g):

Dried fruits:

Potassium	817mg
Calcium	790mg
Manganese	5.1mg
Iron	20.8mg
Copper	1.5mg
Zinc	1.9mg

Source: Advance Journal of Food Science and Technology

Fresh fruits:

Moisture	13%
Fiber	20.5%
Ash	5.2%
Protein	7.7%
Lipids	1.7%
Carbohydrates	66%
Reducing Sugar	13.8%
Starch	44.4%

Related Species:

Grewia flava is a very similar shrub with regard to appearance and function. *G. flava* is used in much the same way as *G. tenax*. *Grewia flava* seems to tolerate cooler climates (such as Mediterranean climates) and heavier soils more than *G. tenax*. Other similar species include *G. flavescens* and *G. villosa*. All of these species produce edible fruit.

Zabala

Scientific name: *Lardizabala biternata*
Family: Lardizabalaceae
Synonyms: *Lardizabala funaria*; Zabala Fruit, Aquiboquil, Cogüilera
Zone: 8b

Figure 60: *L. biternata*. Santa Barbara, California. Credit: Mark Viergutz

Overview:

Zabala is an evergreen vine or liana that can grow up to 10 meters (33ft) and develop a trunk up to 20cm thick. The leaves are usually trifoliate, leathery, and dark green. The plant is usually monoecious with separate male and female flowers, but sometimes it can be dioecious. In any case, the plants are self-infertile, and require the presence of another genetically distinct individual in order to cross-pollinate and set fruit. Zabala vines bloom at the beginning of the autumn and give rise to 5cm long, cylindrical fruits, which contain a sweet pulp.

Habitat and Requirements:

Zabala is native to central – southern Chile from Valparaíso to the Los Ríos region. Thus the plant experiences a very wide rainfall range of between 340-2400mm of average annual precipitation. This range is dominated by a Mediterranean climate, especially in the north. Even in the south, however, there are markedly drier conditions in the summer than during the rest of the year.

The plant is shade-tolerant, and is very well-adapted to growing in areas of deep shade such as in the understory of large trees in woodlands on south-facing slopes (north-facing in the Northern Hemisphere). The plant can tolerate full sun, but generally prefers areas with at least some protection from direct sunlight. The plant prefers well-drained soil, but is generally tolerant of most soil conditions as long as there is ample mulch and/or leaf litter.

Uses:

Fruit and Seeds:

The delicious fruits are somewhat similar to passion fruits and can be eaten raw or cooked. They are very popular in the markets of Chile.

Miscellaneous uses:

The stem fiber is very useful in crafts such as basketmaking.

Propagation:

Propagation with this species is difficult. Seeds should be cold stratified in a refrigerator for two months before sowing in flats. Even after cold stratification, the seeds generally take from 10-15 months to germinate. Plants can also be propagated from cuttings of half-ripe wood. Cuttings should be 5-10cm long and have a heel. The cuttings are best taken during the summer. It is important that young plants and cuttings avoid full sun until they are well established.

Goji Berry

Scientific name: *Lycium barbarum*
Family: Solanaceae
Synonyms: Wolf Berry, Chinese Matrimony Vine, Bocksdorn
Zone: 6b

Overview:

Goji Berry is a deciduous, sprawling shrub growing up to 2.5m (8ft) in height with a 4m (13ft) lateral spread. The plants are self-fertile, and produce hermaphroditic, star-shaped, purple flowers in the summer, which give rise to small, bright red berries in the fall. These berries have become very popular recently due to their health benefits, and are widely marketed, which has caused demand to outstrip supply in many cases. Goji Berries are a proven crop for dry, temperate regions.

Habitat and Requirements:

The exact origin of Goji Berry plants is obscure, but many believe the province of Ningxia, located in central China, is the origin of the species most commonly marketed today. There is little certainty regarding the plants' native habitat, but many sources suggest that Goji Berries are riparian. Goji Berries are cultivated, and have naturalized in the Ningxia province of China, which receives from 190-700mm of average annual rainfall. In the Ningxia province, large scale plantings of Goji Berries are being carried out in order to help the region resist desertification caused by the spread of the three deserts that surround the province.

Goji Berry shrubs are well adapted to growing in sandy deserts, because of their root systems that vigorously grip the soil and help prevent erosion. Plants are very adaptable, tolerating acid and alkaline soils, and, although they are native to an area with a continental climate, they also tolerate seaside conditions, suggesting they are mildly salt-tolerant and wind-resistant. They are known to tolerate extremely hot temperatures up to 46°C (115°F) and can grow just as well in frost-free or nearly frost-free temperate regions, such as Mediterranean climates. The plants prefer full sun, but are also tolerant of partial shade.

Uses:

Fruit and Seeds:
The fruits are bright red, highly nutritious, sweet and have become the focal point of many products at health food stores. Fresh fruits do not keep, so fruits are usually dried in food dehydrators when dealing with large quantities or when shipping.

Leaves:
The leaves can be eaten raw, cooked or made into a tea, but caution should be taken as they may contain small amounts of atropine or other toxic nightshade alkaloids. Cooking the leaves helps decrease the risk of adverse effects. In any case, there is a long history of human use without adverse effects. The young shoots are consumed in the same ways.

Miscellaneous uses:
Goji Berry shrubs are a useful dune-fixing plant, because they help keep loose soil from eroding due to wind or water. Similarly, the plants are also important for the stabilization of river banks. Goji Berry shrubs can be coppiced and heavily pruned to keep them at a desired height.

Propagation:

Propagation is very easy. In the spring, the small seeds should be soaked in water overnight and then planted at a depth of about 5mm. Most seeds will germinate within one month. The seedlings are easy to transplant. In colder areas, the plants should be protected during their first year.

Plants can also be propagated easily from cuttings. Hardwood cuttings of the newest growth should be taken during the winter, and placed on bottom heat. Perhaps the easiest propagation technique is the division of the root suckers, which the plants often readily produce. In winter the root suckers may be dug out, bare-rooted, and potted up or planted elsewhere. Goji Berry plants also form adventitious roots wherever the stems touch the ground, which makes layering another easy propagation method.

Potential Drawbacks:

Figure 61: *L. barbarum*. Zhongning, Ningxia, China
Credit: Paul Gross & Richard Zhang

Under ideal conditions, the plants can grow vigorously, and become invasive. Birds are often attracted to the fruits, and can disperse the seeds to unwanted areas. Pruning the plant encourages root suckers. The plants are often spiny.

Nutrition:

Goji Berries are highly nutritious and have been shown to have anti-cancer effects, help lower cholesterol, and help prevent skin damage caused by ultraviolet radiation. The nutritional properties of Goji Berries are very well documented.
Dried berries (per 100g):

Energy	82.7kcal
Protein	10.6g
Fat Total	3.6g
Saturated Fat	0.92g
Trans Fat	0g
Total Carbohydrates	22g
Sugars	17.1g
Polysaccharides	49mg
Dietary Fiber	7.4g
Selenium	50mcg
Sodium	22mg
Zinc	2mg
Calcium	112mg
Iron	8.42mg
Vitamin C – ascorbic acid	18,400mg
Beta-carotene	7.45mg
Thiamin (Vitamin B1)	0.16mg
Riboflavin (Vitamin B2)	1.3mg
Zeaxanthin	82.5 to 200mg
Amino acids	8.37mg

Source: Goji berry amount and nutrition facts conversion

Related Species:

Lycium chinense is a genetically distinct species used in the same ways as *L. barbarum*. *L. barbarum* is often erroneously labeled as *L. chi-*

nense, and many sources consider them to be synonyms of the same species. Of the two species, *L. barbarum* is the species that has been the subject of most scientific studies, and has been the most widely produced, marketed, and consumed throughout the world.

Moringa

Scientific name: *Moringa oleifera*
Family: Moringaceae
Synonyms: Drumstick Tree, Horseradish Tree, Miracle Tree, Ben Tree, Benzolive, Malunggay
Zone: 9a (The plants will die back to the ground when subfreezing temperatures occur.)

Overview:

Perhaps the most utilitarian species of this book, Moringa is often seen as an extremely nutritious, multipurpose tree, greatly esteemed by the people who rely on its products. Nearly all parts of the tree are edible, and have other uses as well. These drought-deciduous trees are extremely fast-growing, and, under ideal conditions, can grow up to 2.5m (8ft) in one to three months, and sometimes attain a maximum height of up to 16m (52ft), but generally only about 10-12m (33-39ft). Full size trees have a very sparse canopy and are useful in agroforestry, because they allow filtered sunlight to pass through to the understory in quantities that are very beneficial to growing typical garden crops in hot and dry climates. The yellowish-white flowers appear throughout the year and give rise to very long, slender pods (up to 120cm long), which give the plant the appearance of being from the pea family. Moringa produces tripinnately compound leaves that are shed during periods of prolonged drought. Moringa trees have thick, fleshy taproots that aid the trees in finding and storing water. The trees usually only live for about 20 years.

Habitat and Requirements:

The exact origin of Moringa is unknown, yet most evidence suggests that Moringa is native to the sub-Himalayan, subtropical zone of

Figure 62: *M. oleifera* foliage. Tucson, Arizona

northwest India and a small portion of eastern Pakistan. Most rain in this region falls during the summer monsoon season, and winters tend to be rather dry. Moringa have been widely cultivated throughout the arid and semi-arid subtropics and tropics of the world. The trees grow wherever rainfall exceeds 300mm per year (optimal growth above 500mm per year), and are also adaptable to areas with abundant annual rainfall of up to 4,000mm. Trees have been known to grow in areas that receive less than 300mm of average annual rainfall, however, trees in these areas are almost always found near bodies of water, in riparian corridors, or wherever the water table is high. Moringa trees that grow in drier areas produce sparser foliage, and are less productive.

The trees prefer well-drained loam and clay loam soil, and do not tolerate prolonged waterlogging. Although they prefer neutral to slightly acidic conditions, Moringa are known to tolerate soils with a pH of 4.5 up to at least 8.5. Due to their quick growth rate, Moringa are often grown as annual crops in colder, temperate climates as long as these temperate areas have hot summers. Moringa tolerate extremely hot temperatures, and require full sun.

Uses:

Fruit and Seeds:

The pods are cooked like green beans and eaten. They have a taste similar to asparagus. Under ideal conditions, a single tree can produce up to 70kg of pods in one year. The seeds are also edible and contain about 35% oil, which can be used for soap-making. The oil is similar to olive oil, has a sweet flavor, resists rancidity very well, and is often used in cooking. The seeds can also be cracked to expose their kernels, which are effective in the water purification process, since they have coagulant and antimicrobial properties. They are smashed and added to turbid water at the rate of 1 or 2 seed kernels per liter of water. The seed kernels bind with the particles in the water and cause them to sink. This leaves relatively clean water.

Flowers:

The flowers are eaten raw or cooked, and are also a good source of nectar for bees.

Leaves:

The leaves are eaten raw or cooked. In order for trees to produce dense foliage, they must be pruned regularly to be kept as a hedge at about head height. The foliage can also be used as fodder for livestock.

Roots:

The roots can be eaten, but are generally consumed as a condiment because of their powerful, horseradish-like taste. The root bark contains toxic alkaloids, however, and must be removed from the root before consumption. Other alkaloids are likely present in the roots, so they should not be consumed regularly.

Wood:

The wood is light and unacceptable for construction, but can be used for its pulp in paper production. A blue dye can be obtained from the wood.

Bark:

A fiber that can be used to make mats and small ropes is obtained by smashing the bark. The bark can also be used for tanning hides.

Miscellaneous uses:

Figure 63: *M. oleifera* flower. Tucson, Arizona

Moringa can be planted closely and maintained as a hedge. Since trees are fast-growing and can be readily coppiced and pollarded, they provide a reliable source of mulch.

Propagation:

Moringa trees are easy to propagate. Seeds lose their viability over time, but can remain viable for several years if stored in cool (3°C), dry conditions. Seeds require no pretreatment, but may be soaked overnight in water before planting. Seeds should be planted at a depth of 2cm. When fresh, seeds have great germination rates, and most will germinate within two weeks. Plants are so fast-growing, that from seed, a tree can attain a height of 5m in one year.

Moringa can also be shield budded and propagated by cuttings. Stem cuttings are very easy to root. Also, truncheons 4-5cm in diameter and up to 1m in length can be planted directly into the ground and will usually root.

Potential Drawbacks:

The wood is rather fragile and can break when subjected to high winds. Aside from the roots, the pods in some varieties are bitter and reportedly toxic when consumed in large quantities. Root rot is likely in heavy soil when wet and cold. In general, the plants should avoid heavy clay soil.

Nutrition:

Moringa are very nutritious trees, and have been widely utilized by NGOs and other aid organizations to combat malnutrition in impoverished tropical and subtropical regions of the world.

Leaves (per 100g):

Water	78.66g
Energy	64kcal
Protein	9.40g
Total lipid (fat)	1.40g
Carbohydrate, by difference	8.28g
Fiber, total dietary	2.0g

Calcium	185mg
Iron	4mg
Magnesium	147mg
Phosphorus	112mg
Potassium	337mg
Sodium	9mg
Zinc	.60mg

Vitamin C	51.7mg
Thiamin	0.257mg
Riboflavin	0.660mg
Niacin	2.220mg
Vitamin B-6	1.2mg
Folate, DFE	40mcg_RAE
Vitamin A, IU	7564IU

Pods (per 100g):

Water	88.20g
Energy	37kcal
Protein	2.10g
Total lipid (fat)	.20g
Carbohydrate, by difference	8.53g
Fiber, total dietary	3.2g

Calcium	30mg
Iron	.36mg
Magnesium	45mg
Phosphorus	50mg
Potassium	461mg
Sodium	42mg
Zinc	.45mg

Vitamin C	141mg
Thiamin	.053mg
Riboflavin	.074mg
Niacin	.620mg
Vitamin B-6	0.120mg
Folate, DFE	44mcg_DFE
Vitamin A, RAE	4mcg_RAE
Vitamin A, IU	74IU

Fatty acids, total saturated	.033g
Fatty acids, total monounsaturated	.102g
Fatty acids, total polyunsaturated	.003g

Source: USDA National Nutrient Database

Related Species:

Moringa stenopetala is a very closely related species, native to Ethiopia and northern Kenya that is used in the same ways as *Moringa oleifera*, but is more effective at purifying water and is more drought-tolerant.

Lucumillo

Scientific name: *Myrcianthes coquimbensis*
Family: Myrtaceae
Synonyms: *Reichea coquimbensis*; Lucumilla, Arrayán, Arrayán de La Serena
Zone: 10a

Overview:

Lucumillo is a small, evergreen shrub growing up to 1.5m (5ft) tall with a lateral spread of 2-3m (6.5-10ft). The plant has aromatic foliage. The plant is self-fertile, and produces hermaphroditic, white flowers, which give rise to fleshy fruit that turn red when ripe. The fruits contain 1-2 seeds each.

Habitat and Requirements:

Lucumillo is an endangered species located along a 66km strip of land on the coast of the Coquimbo province of Chile from 29-30° south latitude. Lucumillo plants grow naturally along the coast, and are rarely found more than 2km inland. Its native habitat is a Mediterranean climate that receives between 100-200mm of average annual rainfall, which falls almost exclusively during the winter months. During the summer months the region commonly receives morning fog, which sometimes entails mist or light drizzling. This fog usually burns off by the early afternoon.

The plants require full sun, but tolerate a little shade when they are young. They grow in areas of rocky, granite outcroppings, and are typically found on poor, rocky soils.

Figure 64: *M. coquimbensis*. Coquimbo, Chile.
Credit: Leslie Landrum. ASU Herbarium

Uses:

Fruit and Seeds:

The plant produces small, fleshy, red fruits that can be eaten raw or cooked.

Leaves:

In Chile goats are reported as being a key contributor to the decline of Lucumillo plants. This suggests that the plants may be safely browsed by goats, and perhaps other animals.

Propagation:

Seeds should be soaked in water for 1 or 2 days before planting in flats. Additionally, seeds can be pretreated by nicking them or by soaking overnight in an acid solution. The seed flat should be misted frequently to keep them moist. The seeds can take a long time to germinate. Vegetative propagation is likely possible via stem cuttings, but is seemingly unexplored.

Potential Drawbacks:

Lucumillo is relatively unknown outside of its native habitat. Very little research has been done to determine the plant's cultivation requirements.

Myrtle

Scientific name: *Myrtus communis*
Family: Myrtaceae
Synonyms: Common Myrtle, Classic Myrtle, True Myrtle
Zone: 8b

Overview:

Myrtle is an evergreen shrub or small tree growing up to 4.5m (15ft) in height with a lateral spread of up to 3m (10ft). The trunks on mature specimens have peeling old bark, which reveals smooth, reddish-brown new bark. Myrtle flowers in late summer, producing beautiful, white flowers. The flowers are hermaphrodite and plants are self-fertile. Depending on the individual specimen, the flowers either give rise to yellowish, light-colored, small (about 8mm in diameter) fruit, or to dark blue fruit of the same size.

Figure 65: *M. communis* fruit. Superior, Arizona

Habitat and Requirements:

Myrtle is native to the Mediterranean basin, and is predominantly found in coastal areas. As such, the tree withstands maritime exposure, and is well adapted to warmer, temperate parts of the world that receive nearly all their rainfall during the winter. Nonetheless, the plants are also adaptable to hot deserts, and grow well in regions with summer rain. The plants are well adapted to growing in areas that receive roughly 200-800mm of average annual rainfall.

Myrtle prefers to grow in well-drained soil and tolerates a wide pH range, but generally prefers neutral to slightly alkaline soils. The plants require full sun.

Uses:

Fruit and Seeds:

The fruits can be eaten raw or cooked and are best used as an aromatic spice. The fruits make an excellent food flavoring and can help improve digestion.

Flowers:
The flowers and flower buds can be eaten, and have a sweet flavor. These make a great addition to salads.
Leaves:
The foliage of the plant is heavily aromatic due to the high concentration of essential oil found in the leaves. Essential oil is found throughout the entire plant, but the highest concentrations are found in the new growth of stems and leaves. This essential oil has antiseptic properties and can be extracted from the plant at the rate of 10g of oil per 100kg of leaves. This oil can be eaten as a condiment, and the whole leaves can also be used as a spice. The oil is also used in beauty products, soap, and perfume.
Wood:
The hard wood is very fine-grained and individual branches can be used as tool handles. Larger pieces of wood can be used for furniture and other structures. A high quality charcoal can also be produced from the wood.
Miscellaneous uses:
The plants tolerate frequent pruning and can be maintained as hedges.

Propagation:

Myrtle can be propagated from seed and from stem cuttings. Seeds should be soaked for 24 hours in warm water, before being planted into a seed flat in the spring. The seeds should be covered slightly with soil so that the seed is buried at a depth roughly equal to its diameter. The seeds can take up to a few months to germinate.

Heel cuttings should be taken in late fall. The cuttings are taken from the current year's growth. The cuttings should be rooted with bottom heat in a frost free area, absent of direct sunlight. The plant has a relatively quick growth rate when young, but slows down with age.

Nutrition:

Fruits:	
Crude Oil	2.37%
Crude Protein	4.17%
Crude Fiber	17.41%
Crude Energy	11.21kcal/g
Reducing Sugar	8.64%
Tannin	76.11mg/100g
Ash	0.725%
Water-Soluble Extract	52.94%
Essential Oil	0.01%

Source: Journal of Food Engineering

Figure 66: *M. communis* foliage. Superior, Arizona.

Related Species:

Myrtus nivellei is a similar species, found up to 1000m in the mountains in the middle of the Sahara Desert. Its native habitat is an area that receives 50-100mm of rainfall per year. The plant's leaves are drunk as a tea for liver health.

Transvaal Kooboo-Berry

Scientific name: *Mystroxylon aethiopicum subsp. aethiopicum*
Family: Celastraceae
Zone: 9b

Overview:

Transvaal Kooboo-Berry is the name given to the three subspecies of *Mystroxylon aethiopicum*. In this case, the name is used specifically for *Mystroxylon aethiopicum ssp. aethiopicum*, which is a medium size, evergreen tree attaining heights of up to 15m (49ft). It produces small, yellowish-green flowers in clusters starting in early spring and lasting until the end of the summer. The flowers give rise to yellowish green fruits that turn red when fully ripe, which is usually between late fall and early winter.

Habitat and Requirements:

The Transvaal Kooboo-Berry is a plant native to the southern parts of the Western and Eastern Cape provinces of South Africa. The trees are found growing in a wide variety of habitats including riparian zones as well as the rocky ridges of hillsides. The plants grow well in areas that receive between 350-1200mm of average annual rainfall. The plants can tolerate partial shade when they are very young, but generally require full sun.

Uses:

Fruit and Seeds:

The fleshy, ripe, red fruits are sweet, and enjoyed by humans and animals alike.

Leaves:

Younger foliage is readily eaten by cattle and goats.

Flowers:

The trees are excellent for apiculture.

Wood:

The etymology of this plant's scientific name stems from the Greek mystro meaning spoon and –xylon meaning wood. The wood of the tree is hard and tough, and, aside from being used to make wooden

Figure 67: *M. aethiopicum subsp. schlechteri*. Piti, Mozambique
Credit: Geoff Nichols

spoons, it is also an excellent wood for making tool handles such as axes, shovels and picks.

Bark:

A preparation of the bark can be used as a vermifuge for cattle. For humans, it can be used to treat stomach aches, coughs and anemia. The bark also produces a brown dye that is useful for tanning.

Miscellaneous uses:

Since the trees are rather dense, they can be planted as effective noise barriers and windbreaks.

Propagation:

All fruit flesh should be washed off of the seeds before planting, because it contains growth inhibiting properties that prevent seeds from germinating and/or developing. Seeds can be soaked in water for 1-2 days and then planted into seed flats. Most seeds should germinate within one month. When they develop true leaves, they can be potted up into larger containers. The plants should be grown in partial shade for the first 2-3 months of their life, and can then be acclimated to full sun.

Cuttings can also be taken from the current year's growth, and should be taken at the tips of branches. The cuttings should be placed in sand or perlite, subjected to bottom heat, and kept out of direct sunlight until they root.

Potential Drawbacks:

Although the fruits are usually sweet, certain specimens may produce sour and/or bitter fruit.

Related Species:

The other subspecies besides *ssp. aethiopicum* are *ssp. schlechteri* and *ssp. burkeanum*. *Ssp. schlechteri* is the most widespread of all the species and grows from southern South Africa up through Eastern Africa to the northeastern portion of the continent. The tree also occurs on many of the islands off the east coast of Africa, including Madagascar. *Ssp. burkeanum* has a limited range, growing in dry, rocky areas of the Highveld of northern South Africa. All subspecies produce edible fruit, and have similar uses.

Tortoise Berry

Scientific name: *Nylandtia spinosa*
Family: Polygalaceae
Synonym: Skilpadbessie
Zone: 8a

Figure 68: *N. spinosa*. Bloemfontein Botanical Gardens, South Africa
Credit: Adrian de Wit

Overview:

Tortoise berry is a very ornamental, small, woody shrub growing up to 1m (3.2ft) tall with an equal spread. It has small, evergreen leaves born on spiny stems. Starting in the fall, and lasting usually until spring, the plants produce brilliant, pink to purple flowers. During their peak blooming period, individual plants are blanketed with these flowers, producing a spectacular display. The flowers look similar to small pea flowers, and give rise to bright red fleshy fruits during the summer months.

Habitat and Requirements:

Tortoise Berry is native throughout western and southern South Africa from Namaqualand, through Western Cape, and into the southern portion of the Eastern Cape. As such, Tortoise Berry is widely adaptable and tolerates a broad range of climatic conditions. The plants can grow in mountainous regions and are also very well adapted to growing along the coast. The shrubs grow well in areas that receive between 200-1200mm of average annual rainfall. Tortoise Berry grows just as well in Mediterranean climates as it does in areas that receive summer rain.

The shrubs prefer to grow in sandy or rocky, well-drained soils, and may be stunted or suffer disease problems in heavy soils. Aside from growing inland, plants grow well in coastal areas and areas immediately adjacent to the ocean, suggesting that they not only tolerate high winds, but also salt spray, and relatively salty soil. The plants require full sun.

Uses:

Fruit and Seeds:

The shrub received its name because tortoises enjoy eating the fruits. Humans can also eat the fruits. They have a thirst-quenching taste and are somewhat astringent, but are sweet enough to be enjoyed by adults and children alike.

Leaves:

The leaves and young stems can be made into a tea in order to treat colds, abdominal pain, tuberculosis, and to help digestion. Mice studies show that the leaves also contain anticonvulsant properties that may be used to treat epilepsy.

Miscellaneous uses:

The plants can be planted to stabilize sandy areas, helping to prevent erosion.

Propagation:

Cuttings are very difficult to root, so propagation from seed is the preferred method. The seeds can be soaked in water overnight, and then sown into seed flats. Most seeds will germinate within one month.

When the seedlings are a couple months old, they can be transplanted from the flats. Care should be taken when transplanting, because the seedlings develop taproots up to three or four times as long as the height of the seedling. The seedlings should be potted up and left to grow for up to 1.5 years in the nursery, because they are slow to establish themselves, and are very fragile when young.

Potential Drawbacks:

The growth rate can be slow when young, and in areas of heavy soil.

Nutrition:

The fruits are rich in Vitamin C.

Related Species:

Nylandtia scoparia is the only other plant in the genus. This plant also produces small, edible fruit. It has nearly the same range as *N. spinosa*, and the plants often grow together in the wild. *N. scoparia* has a more restricted distribution range, however, growing only in the Western Cape from the Bokkeveld Mountains to the town of Darling.

Markouba Grass

Scientific name: *Panicum turgidum*
Family: Poaceae
Synonyms: Merkba, Taman, Guinchi, Du-ghasi, Afezu
Zone: 10a

Overview:

Markouba Grass is a halophytic, perennial bunchgrass up to 1m (3.2ft) tall that can be used as fodder, as a pioneer plant, and as a source of food where no other plants will grow.

Habitat and Requirements:

Markouba Grass is perhaps the most drought-tolerant potential crop on earth. It is native to the Sahara desert, eastward through the Middle East and into Pakistan. Markouba Grass is known for its

Figure 69: *P. turgidum.* Tagant Plateau, Mauritania
Credit: Unknown

ability to colonize sand dunes in the middle of the Sahara. It grows in areas that receive as little as 25mm of annual precipitation. The maximum annual rainfall tolerance of the plant appears to be about 400mm of rainfall.

The plant colonizes sandy areas via its root exudates, which are a form of glue that actually trap sand particles and residual moisture.

Uses:

Fruit and Seeds:
In the Sahara the grains are commonly ground into edible flour and are a staple food.

Leaves:
The leaves can be used for thatch, and are also an acceptable animal fodder.

Miscellaneous uses:
The grass is useful in dune reclamation, and is often used as a pioneer

plant for other species to establish themselves. Indeed, *Acacia tortilis var. raddiana* is often planted in the center of a Markouba Grass bunch so that it has a microclimate to establish itself.

Propagation:

With enough water, seeds will germinate readily on sandy soil. Otherwise, divisions are an easy way to propagate the plants. Air pruning containers should be used in the nursery because traditional containers may foster too much humidity around the root zone.

Jacket Plum

Scientific name: *Pappea capensis*
Family: Sapindaceae
Synonyms: Indaba Tree, Bushveld Cherry, Gulaswimbi
Zone: 8b

Overview:

The Jacket Plum is a long-lived, evergreen tree, which reaches heights of 2-8m (6.5-26ft). The tree has a slow to moderate growth rate. The leaves look similar to those of oak trees, especially when young. The tree produces pinkish-bronze new leaves that have sharply-toothed margins. In its native habitat, Jacket Plum flowers between the months of September to May (March to November in the Northern Hemisphere) and produces fruit between the months of December to July (June to January in the Northern Hemisphere). The flowers are greenish catkins, which are followed by green fruits. When ripe, the green "jacket" splits open to reveal a bright red, fleshy fruit about 10mm in diameter.

Habitat and Requirements:

Jacket Plum has a very wide, native distribution range. It is found growing throughout South Africa from the Northern Cape throughout the Karoo and Western Cape, to the Eastern Cape, Kwa-Zulu Natal, northern South Africa and northwards throughout tropical southern and eastern Africa as far north as Kenya. As such, the plant is widely

Figure 70: *P. capensis*. Kwandwe Private Game Reserve, South Africa
Credit: Lorraine Solomon

adaptable and grows well from arid to Mediterranean and from subtropical to tropical climates. The plant is well adapted to growing in areas that receive between 200-1200mm of average annual rainfall. The plant grows best in full sun, and tolerates nearly all soil types. The plant grows markedly faster in warm and humid climates, and is rather slow-growing in cold and arid climates.

Uses:

Fruit and Seeds:

The fruits are very juicy and delicious and have a sweet to tart flavor. They can be eaten fresh, made into preserves or fermented to produce an alcoholic beverage. The seeds are also edible and their oil is very useful. It can be used for making soap, restoring hair, and as a cure against ringworm. Furthermore it is useful as an industrial lubricant. As much as 2,400 liters of oil/hectare can be produced, which yields approximately 1,560 liters of biodiesel.

Leaves:
The leaves are often browsed by wild animals, and are also a suitable source of fodder for livestock. The leaves are also said to be effective at killing snails.
Wood:
The wood can be used to make furniture and utensils.
Bark:
The bark (as well as the leaves and seed oil) can be used against baldness, ringworm, eye infections, venereal disease, bloody nose and chest complaints.
Miscellaneous uses:
The plants often have a thick growth habit, making them useful as a plant for windbreaks.

Propagation:

Fresh seeds should be used for propagation. The seeds should be completely cleaned of all the fruit flesh. Nicking seeds and soaking them overnight in water improves germination rates. They are then sown into flats at a depth of 5mm. The seed flats should be kept in light shade until the seedlings germinate. Germination usually occurs after 1-3 months. The seedlings are then potted up when they are between 20-50mm tall. They have somewhat fragile taproots, so care should be taken when potting the seedlings up at this stage. The plants are very slow-growing when young, but they eventually pick up speed as they get older.

Related Species:

The Jacket Plum is the only species within its genus. Previously, the species was erroneously subdivided into two different varieties: *Pappea capensis var. capensis* and *P. capensis var. radlkoferii.* This distinction was made in order to differentiate trees growing in arid climates from trees growing in more humid climates.

Galupa

Scientific name: *Passiflora pinnatistipula*
Family: Passifloraceae
Synonyms: Gulupa, Pasionaria Bejuco, Tintin
Zone: 9a

Overview:

Galupa is an evergreen vine that can climb trees up to heights of 15 meters (49ft). It has 3-lobed leaves and produces beautiful flowers that have hues of pale pink, purple and blue, which give the plant much ornamental value. The flowers are self-fertile, and give rise to green, nearly globe-shaped fruits, which turn yellow when ripe.

Habitat and Requirements:

Galupa is cultivated widely, from Colombia to Chile. Since it has been cultivated so extensively in South America, Galupa's exact origins remain to be proven. Nonetheless, it seems most populations occur throughout the region of southern Peru and eastern Bolivia. Other sources claim the species to be native to a very small part of central Chile.

Such sources claim that the vine grows wild along a few coastal streams of the Valparaíso region, and the extreme south of the Coquimbo region. The rainfall in its Peruvian/Bolivian mountainous habitat ranges from about 450-700mm of average annual rainfall. This rainfall occurs mainly during the summer months. In Chile, the plant's range receives about 150-400mm of average annual rainfall. This rain falls primarily during the winter.

The plant prefers loam soil, and tolerates clay soils as well, as long as they are biologically rich. The plant prefers to grow in areas of partial shade, such as beneath light-canopied trees. Galupa will likely grow well wherever Bolivian Mountain Coconuts (*Parajubaea torallyi var. torallyi*) thrive.

Uses:

Fruit and Seeds:

The ripe fruits are very sweet, and can be used for juices, preserves, or simply cut in half and eaten with a spoon. They taste similar to the typical Brazilian Passionfruit (*Passiflora edulis*).

Propagation:

The plants are easy to propagate by seed. They should be soaked overnight in water and then sown into seed flats, and left in partial shade. Most seeds will germinate within one month. Galupa plants can also be propagated by stem cuttings of the previous year's growth. Young cuttings are much more prone to desiccation. Also, cuttings can be grafted on to *Passiflora mollissima*, in order to produce more vigorously.

Potential Drawbacks:

Galupa is sensitive to solar exposure in hotter, more arid regions, and is relatively intolerant of extreme heat.

Figure 71: *P. pinnatistipula.* Santa Barbara, California
Credit: Mark Viergutz

Nutrition:

The nutritional profile of Galupa is likely similar to that of Purple Passionfruit (*Passiflora edulis*). The nutritional profile for 100g of the pulp of Purple Passionfruit is as follows:

Water	72.93g
Energy	97kcal
Protein	2.20g
Fat	.70g
Carbohydrates	23.38g
Fiber	10.4g
Sugar	11.20g

Calcium	12mg
Iron	1.60mg
Magnesium	29mg
Phosphorus	68mg
Potassium	348mg
Sodium	28mg
Zinc	.10mg

Vitamin C, total ascorbic acid	30mg
Riboflavin	0.130mg
Niacin	1.500
Vitamin B-6	0.100
Folate, DFE	14mcg_DFE
Vitamin A, RAE	64mcg
Vitamin A, IU	1272IU
Vitamin E (alpha-tocopherol)	0.02mg
Vitamin K (phylloquinone)	0.7mcg
Fatty acids, total saturated	0.059g
Fatty acids, total monounsaturated	0.086g
Fatty acids, total polyunsaturated	0.411g

Source: USDA National Nutrient Database

Related Species:

Passiflora mollissima grows in Andean valleys from Venezuela and eastern Colombia to Bolivia and Peru. The plant performs well in Cuzco, Peru and is a high yielding species with up to 300 fruits per vine per year under ideal conditions. In a single year, 30,000kg of fruit can be produced on one hectare of land. *Passiflora tripartita var. mollissima* is a species very similar to *P. mollissima* that is also very vigorous. *P. mandonii* is probably the most similar species to *P. pinnatistipula*, and shares roughly the same distribution range in Peru and Bolivia. Occasionally, a white-flowered variety of *P. pinnatistipula* is found in the wild. This variety is said to be less vigorous.

Single-Leaf Pinyon

Scientific name: *Pinus monophylla*
Family: Pinaceae
Synonym: Single-Leaf Pinyon Pine
Zone: 5a (Hardy to about -30°C or -23°F)

Figure 72: *P. monophylla*. Swall Meadows, California
Credit: Jane Richardson

Overview:

Single-Leaf Pinyons commonly live to be 300-400 years old, and have been known to live up to 800-1000 years. The trees often attain heights of 6-12 meters (20-40 feet). The trees have taproots, but also develop very extensive lateral root systems, which stretch well beyond the canopy of the tree. The scientific name, *monophylla*, means one leaf or single-leaved, and refers to the fact that it is the only pine species known to have a single needle per sheath. The pine is monoecious, and, like all pines, wind-pollinated. The cones develop and ripen during autumn, and contain numerous, 10-17mm long seeds, which are edible. The trees begin bearing when they are about 35 years old, and reach maximum productivity when they are about 160-200 years old.

Habitat and Requirements:

The fact that Single-Leaf Pinyon Pine has a distribution range that receives 200-460mm of average annual precipitation, makes it the most drought-tolerant pine species native to the U.S.A. In many areas, however, the trees receive additional moisture from snow melt. Most of the precipitation falls during the winter months, and summers are hot and dry.

The pines are an upland species, and primarily grow in mountainous areas often on steep, rocky slopes or ridges, but also in alluvial fans. These habitats indicate that the trees require good drainage, and should not be planted in areas with heavy clay soil. Single-Leaf Pinyons are found growing in soil with a pH of 6-8.

Single-Leaf Pinyons are found mostly growing in mountainous areas, and are concentrated most heavily in Nevada and Utah, but are also found in Arizona, California, and northern Baja California.

Uses:

Fruit and Seeds:

The seeds are covered with a thin shell. This shell is cracked open and the "pine nuts" can be eaten raw or cooked. They are most commonly roasted and eaten by humans, but they can also be used as fodder for livestock.

Wood:

The wood is best used for fence posts, and as fuelwood. The wood burns very hot, and can be converted to charcoal to melt silver.

Bark:

The inner bark can also be used as animal fodder.

Miscellaneous uses:

The pitch can be used as an adhesive, as caulking material, and as paint binder.

Propagation:

Since seeds are short-lived, they should be planted fresh. Fresh seeds often yield germination rates of 85-95% but lose much of their viability in one year or less. Before being sown in the spring, the seeds should be cold stratified in a freezer and/or refrigerator for 30-90 days. Care should be taken when transplanting the seedlings, because of their deep taproot. Taproots of up to 15cm (6in) have been measured ten days after germination.

The saplings can be raised in the nursery in deep pots for a couple years before being planted in the field. They establish themselves best in the shade of shrubs, but not under the canopy of other pines. They benefit from the leaf litter and partial shade provided by the shrubs, but must compete for nutrients. Therefore, providing saplings with artificial shade and ample mulch is the quickest way to establish plants out in the field. The trees can probably be grafted in order to select reliable varieties, and to start producing at a younger age.

Potential Drawbacks:

The trees take a long time to start producing and are slow-growing. They only grow about 1 inch per year when they are saplings. The trees are allelopathic, and exude chemicals from their roots which impair the ability of other plants to grow nearby. As such, Single-Leaf Pinyons usually end up forming a monoculture when mature.

Nutrition:

> "Pinyon nuts have outstanding nutritional value. Shelled nuts of singleleaf pinyon have approximately 9.5% protein, 23% fat, and 54% carbohydrate. They are comparable in nutritional value to pecans. They contain all 20 amino acids, with 8 of the 9 essential amino acids are more abundant in pinyon nuts than in amaranth (*Amaranthus spp.*), another important food plant of Great Basin Native Americans. Singleleaf pinyon seeds are especially rich in tryptophan and cystine. They have a large amount of oleic and linoleic unsaturated fatty acids, comprising 42.5% and 42% of the fatty acid total, respectively. Eighty percent of the fatty acids are polyunsaturated. The nuts also provide significant amounts of vitamin A, thiamine, riboflavin, and niacin." Source: Zouhar, Kristin L.

Related Species:

There are three subspecies of Single-Leaf Pinyon. *Pinus monophylla subsp. monophylla* occupies the majority of the range. *Pinus monophylla subsp. californiarum* occupies southern Nevada, southeastern California, and northern Baja. *Pinus monophylla subsp. fallax* is found throughout the southeastern portion of the range from southern Utah to Arizona and into New Mexico.

Pinus edulis is a similar species that grows to the east of *P. monophylla*'s distribution range. Hybrids between *P. edulis* and *P. monophylla* are found where their ranges overlap. *P. edulis* nuts have a higher fat content, and are therefore more palatable.

Other economically important nut pines from similar climates are:

Pinus siberica (Siberian Pine) – This pine is native to Siberia, southeast Russia and into central Mongolia. It is cold hardy down to at least -60°C (-76°F).

Pinus koraiensis (Korean Pine) – This pine is native to eastern Russia, Korea and Central Japan. It is cold hardy down to about -50°C (-58°F). This species has taken the place of *Pinus pinea* as the prominent pine nut of international commerce.

Pinus gerardiana (Chilgoza Pine) – This pine is native to the northwestern Himalayas of Afghanistan, Pakistan and India.

Stone Pine

Scientific name: *Pinus pinea*
Family: Pinaceae
Synonyms: Italian Stone Pine, Umbrella Pine, Pignolia Nut Pine
Zone: 7a

Overview:

The Stone Pine is a large, evergreen tree growing up to 30m (98ft), but is usually seen at a mature height of around 12m (39ft). They attain a rounded canopy, which gives the entire tree the shape of an umbrella, hence the common name Umbrella Pine. The needles are about 6 inches long. The trees usually only live to be about 100 years old, and start producing cones when they are 6-10 years old. The trees are wind pollinated in the spring, and produce cones that ripen in the fall. Each cone holds about 50 seeds, and the seeds make up about 20% of the cone's weight. An annual yield of nuts is about 500kg/hectare at the optimal tree spacing of 10 meters. Pine nuts are a lucrative crop, because world demand far outstrips supply.

Figure 73: *P. pinea*. Santa Barbara, California. Credit: Mark Viergutz

Habitat and Requirements:

The Stone Pine is native to the Mediterranean basin. Therefore, Stone Pine is well adapted to growing in areas that receive between 200-800mm of average annual rainfall, but produce best in areas that receive over 400mm of annual rainfall. In its native habitat the Stone Pine receives the majority of its rainfall during the winter months, and is well-adapted to areas with long, dry, hot summers as well as milder coastal areas. The pine is cold-tolerant, and some sources suggest the trees may survive down to -23°C (-9°F).

The plant is tolerant of coastal conditions such as salty sea spray, and grows well next to beaches in sandy soil. The plant tolerates nearly all soil types including heavy clay, but performs optimally in areas with good drainage, and is most often found growing in rocky soil. The trees prefer a pH range from 5.5-7.5, and are somewhat salt-tolerant. In dry areas, the trees develop strong taproots. In wetter areas, however, the trees fail to develop a taproot and may blow over in heavy winds.

Uses:

Fruit and Seeds:
Most species of pines produce edible nuts, and this species is no exception. The Stone Pine produces perhaps the best-tasting nuts out of all the pine species. The seeds are roughly 20mm x 10mm and enclosed within a woody shell. The cones are harvested with a hook on a long pole when they are brown, and before the scales open. They are then laid out in the sun or heated in an oven in order to open the scales, an act which facilitates seed extraction. Grinding the young pinecones yields a powder that can be used as a food flavoring.

Leaves:
The leaves (needles) yield a green dye.

Resin:
Like all pines, Stone Pine resin contains turpentine and has antiseptic, diuretic, rubefacient, and vermifuge properties. It can be used externally to help heal wounds, burns, sores and boils. Internally, it is useful for treating kidney and bladder complaints as well as to treat rheumatic problems. It can also be used in a steam bath to treat respiratory problems.

Wood:
The wood can be used in general construction and for furniture making; however, the trees should be pruned to a single trunk for timber production. The wood can also be used as a source of fuel.

Miscellaneous uses:
For nut production the trees are planted at 10m (33ft) intervals, but they should be planted at half that distance in order to form windbreaks. The pines tolerate pruning, and can be maintained as a hedge or barrier. The pines can also be used for dune-fixing, since their root system helps prevent erosion of loose soils.

Stone Pines often associate and form a symbiotic relationship with the ectomycorrhizal fungi, *Tuber borchii*, a species of white truffle whose fruit (mushroom) sells for about 400 euros/kg on the world market. The roots of Stone Pine seedlings can be intentionally infected with *Tuber borchii* in order to produce truffles, which naturalize themselves amongst Stone Pine plantations. Since many of the truffles do not fruit above ground, they can be located using well-trained dogs or pigs.

Tuber borchii is well adapted to Mediterranean climates, and fruits best in sandy, coastal areas that receive 600-1600mm of annual rainfall, and where soil pH is 5.2-8. It prefers neutral to slightly alkaline soils, and produces mushrooms in the northern Hemisphere from January to the end of April. The Truffles should start producing after four years from the establishment of the plantation, and may produce up to 15kg per hectare per year when the plantation is more mature.

Propagation:

The seeds require no special pre-treatment, but can be soaked in water for 12-24 hours before planting. They germinate when temperatures are between 10-25°C (50-77°F). They should not be overwatered, since the seeds and young seedlings may rot. The seedlings develop more rot resistance when they are about 10cm tall. Germinated seedlings can be transplanted within two weeks. They should be transplanted to deep pots in order to accommodate their taproot. Care should be taken when transplanting to avoid damaging their taproot. The plants do not like to be in the nursery very long, and should be planted before they reach a height of 90cm. Ideally, seeds are collected in the fall, germinated, raised in the nursery, and planted the following autumn in order to benefit from establishing themselves in the ground during the winter rainy season.

Stone Pines can be grafted in order to select superior cultivars. In most cases they can be grafted onto *Pinus pinea* rootstock, but they can also be grafted on to other pines for quicker growth. Grafting onto *P. radiata* rootstock, for example, can be done with up to an 85% success rate. The best graft is a basal cleft graft, which should be executed in mid-summer. Grafting of nut pines is difficult, and requires precision. The rootstock should be at least two years old, because grafting younger rootstock often results in a delayed mortality, which may not reveal itself until about two years into the future.

Potential Drawbacks:

As mentioned already, in wet areas the trees fail to develop a taproot and may blow over in heavy winds, since they are top heavy.

Nutrition:

Dried pine nuts (per 100g):

Water	2.28g
Energy	673kcal
Protein	13.69g
Fat	68.37g
Carbohydrates	13.08g
Fiber	3.7g
Sugar	3.59g

Calcium	16mg
Iron	5.53mg
Magnesium	251mg
Phosphorus	575mg
Potassium	597mg
Sodium	2mg
Zinc	6.45mg

Vitamin C, total ascorbic acid	0.8mg
Thiamin	0.364mg
Riboflavin	0.227mg
Niacin	4.387mg
Vitamin B-6	0.094mg
Folate, DFE	34mcg_DFE
Vitamin A, RAE	1mcg_RAE
Vitamin A, IU	29IU
Vitamin E (alpha-tocopherol)	9.33mg
Vitamin K (phylloquinone)	53.9mcg
Fatty acids, total saturated	4.899g
Fatty acids, total monounsaturated	18.764g
Fatty acids, total polyunsaturated	34.071g

Source: USDA National Nutrient Database

Related Species:

The 'Fragilis' variety is the most widely cultivated variety because of its thin-shelled seeds that are easier to crack. *Pinus halepensis* is another Mediterranean pine harvested for its edible seeds.

Pinus sabiniana and *P. coulteri* are native to Mediterranean climate regions of California and produce large cones. The latter produces the largest cone of any pine species. Both have edible seeds, but do not taste as good as those of *P. pinea*, due to their lower oil content.

Lucuma

Scientific name: *Pouteria lucuma*
Family: Sapotaceae
Synonyms: *Pouteria obovata*; Lucmo, Lucma, Rucma, Mamón
Zone: 9b

Overview:

Lucuma is an evergreen tree attaining heights of up to 15m (49ft) when mature. Under cultivation, however, the plants are usually pruned shorter. Lucuma has dark-green, leathery leaves and produces round, green-skinned fruits that contain a yellow pulp with usually one or two (sometimes up to 5) large, smooth seeds. The seeds are not edible, but the rest of the fruit is very nutritious. A single mature tree can produce about 500 fruits per year, and individual fruits may weigh up to 1kg. Lucuma trees produce year-round. The trees take 6-10 years to start producing from seed.

Habitat and Requirements:

Lucuma is native to the Andes of Colombia, Ecuador, Peru, Bolivia and Chile. Various strains of the species appear to exist, because many fruits are cultivated close to sea level in northern Peru and central Chile, whereas others are cultivated in the Andean highlands between 1,500 and 3,000 meters (4,920-9,842ft). The Sacred Valley in Peru is home to some of the highest-altitude Lucumas, which are adapted to growing in areas with as little as 600mm of average annual rainfall. Most of the Andean Lucumas, however, are found more commonly at lower elevations where they can grow in microclimates that receive up to 3,000mm of average annual rainfall. The plants are deep-rooted and are able to tap groundwater, an adaptation which allows them to survive droughts lasting many months.

Most of the rain the plants receive occurs during the summer. Nonetheless, the plants appear to be well-adapted to growing in Mediterranean climates, since Chile has developed varieties that grow successfully down to 33° south latitude. The plants prefer a rather mild climate, where the high temperatures are roughly the same year round. Surprisingly, however, the plants tolerate extreme heat, but require considerable irrigation to keep them from withering.

Lucuma prefers deep, alluvial soils that are rich with organic matter. Since the plants require good drainage, they grow well in sandy or rocky soils, but do not grow very well in clay soil, and do not tolerate waterlogging. The trees tolerate moderate salinity, calcareous soil, and are well-adapted to growing in iron-deficient sites. Trees prefer soil with a pH of 6.1-7.5.

Uses:

Fruit and Seeds:

Lucuma fruit are usually lumped into two broad categories: soft (lucuma suave) and hard (lucuma de palo). Hard fruits often have the consistency of the flesh of pumpkins, but are generally more flavorful. Soft fruits, on the other hand, have the consistency of roasted sweet potatoes, and are easier to eat out of hand. The most sought-after fruits are those that have the texture and taste of roasted sweet potatoes drenched in maple syrup. Fruits are most commonly used to make ice cream. Inferior fruits are often used as animal fodder.

When the fruits fall from the trees they are still unripe, and require a couple weeks in a dry, cool area to ripen. The fruits are often dried, and then ground into a flour, which has a considerably long shelf life. Although the seeds are not eaten, they contain high amounts of fatty acids, and are useful for topically healing wounds, because they help accelerate skin regeneration.

Wood:

The wood has a pale color and is very dense and durable. Therefore, the wood is highly valued for its use in construction.

Propagation:

Lucuma is propagated from seed. The seeds require no special pretreatment, but appear to lose their viability over time. The seeds

Figure 74: *P. lucuma*. Santa Barbara, California. Credit: Mark Viergutz

should be placed in moist peat moss, and then sealed in bags. Most seeds will germinate within three months, but other seeds may take up to one year to germinate. The seedlings are easy to handle and pot up. They should be grown in the nursery for at least the first two years of their life. Grafting is feasible, but is rather difficult.

Potential Drawbacks:

As mentioned already, Lucuma should not be grown in areas with soggy, clay soil.

Nutrition:

The fruits are filling, and, unlike most fruits, contain many calories. The fruits have anti-hyperglycemic properties and are potentially useful in preventing or treating diabetes. Fruit pulp (per 100g):

Water	72.3g
Energetic Value	99 calories
Protein	1.5g
Fiber	1.3g
Glucose	8.4g
Fructose	4.7g
Sucrose	1.7g
Inositol	0.06g
Lipids	0.5g
Ash	0.7g
Calcium	16mg
Phosphorus	26mg
Iron	0.4mg
Carotene	2.3mg
Thiamine	0.01mg
Niacin	1.96mg
Ascorbic acid	2.2mg
Riboflavin	0.14mg

Source: family-content.com

Related Species:

Pouteria splendens is a shrub indigenous to coastal, central Chile. The shrub produces bright red, edible fruits.

Desert Quandong

Scientific name: *Santalum acuminatum*
Family: Santalaceae
Synonyms: Quandong, Wild Peach, Sweet Quandong
Zone: 9b

Overview:

The Desert Quandong is an evergreen shrub or small tree that can reach a maximum height of 6m (20ft) with a spread of 4m (13ft). Desert Quandongs usually remain shorter, however, and grow as a sprawling shrub. The plants produce hermaphroditic flowers in late summer followed by fruits that ripen in the spring. These shiny, bright red fruits grow from 15-25mm, have become very popular in Australia as a native "bush food" and have recently begun to be cultivated. Desert Quandong plants are hemiparasites, producing their own sugars via photosynthesis, while extracting many essential elements from the xylem of other nearby plant roots.

Habitat and Requirements:

Desert Quandongs are native to arid and semi-arid Australia (mostly to the lower half of the country), growing across a large portion of the country. They are well adapted to Mediterranean climate regions, as well as hot desert regions and are found where there is between 150-750mm of average annual rainfall. Since Desert Quandongs are found across many regions of Australia, they are adapted to growing in areas where most rainfall occurs in the winter, summer, or year-round. They are less dependent on rainfall, because they attain most of their water from the roots of their host plants.

Desert Quandong plants often parasitize more than one plant at a time, and have the ability to interface with trees, shrubs, and even grasses. Since their root system may spread up to 10m (33ft) in length, the Desert Quandong plants may potentially parasitize any plant growing within a 10m (33ft) radius of the trunk. Although they have the ability to parasitize many different species, they prefer nitrogen-fixing trees and shrubs such as *Acacia spp.* and *Casuarina spp.* as their hosts.

Figure 75: *S. acuminatum*. Badjaling, Western Australia
Credit: Phil Lewis

Desert Quandongs grow well in partial shade when young, but prefer full sun when they are mature. The plants prefer well-drained, sandy soil, but also tolerate loams and clays as long as they are not subject to waterlogging. The plants tolerate salty, coastal winds, and can withstand soil salinity up to 16 deciSiemens per meter (dS/m).

Uses:

Fruit and Seeds:

Mature plants produce between 10-25kg of fruit per year. The kernel comprises about 40% of the fruit's weight. The flesh of the fruit is about 3-5mm thick, and has a tart taste similar to a cross between apricots, peaches and rhubarb. It is easily dried, and can be stored for long periods of time. More commonly, however, the flesh is used to make jams and chutneys.

The rest of the fruit is composed of a large, woody pit that can be cracked open to reveal an edible kernel that contains over 50% fat. The kernel can be eaten raw, but is usually roasted. Due to its oil content, the seed is also highly flammable, and can be burned in a similar way as a candlenut.

Leaves:

The foliage is a favorite of browsing animals, and can be used as fodder for livestock.

Wood:

The wood also contains flammable oils, making it useful for starting fires. The wood is also very heavy and hard, and is close-grained, contributing to its use in furniture making.

Propagation:

Due to their thick seedcoat, the seeds are difficult to germinate. Furthermore, the seeds have a dormancy period, which encumbers the germination of fresh seeds. Optimal germination rates of 35% can be obtained after three years. Generally, however, after one year, seeds demonstrate up to a 30% germination rate. The seed coat should be carefully cracked or nicked. To speed up germination, the seeds can be completely removed from their seed coats, but this method often results in rotten seeds.

A simpler method is to dig small basins near a host plant, and to fill

Figure 76: *S. acuminatum*. Nambung National Park, Western Australia
Credit: Leif Stridvall

the basins with leaf litter and other mulch produced by the host plant. Kept moist, the seeds should germinate within four months. Since the optimal temperature for germination is 15°C (60°F), the plants should be sown at the beginning of winter. In the nursery, seedlings can be kept in containers for at least a year, but should be planted out sooner with a host, since their growth is rather dependent on other plants.

Desert Quandongs can be grafted, and a few reliable cultivars have been selected. These cultivars can be found under the names: Powell's Red Supreme, Saltbush Lane, Powell's Number One, Frahn's Paringa Gem, Nonning, and Nectarbrook.

Potential Drawbacks:

One study showed that when *Melia azedarach* (White Cedar) hosted Desert Quandongs, the plants acquired insecticidal compounds from their host tree, which accumulated to some degree in the fruits. Therefore, caution should be taken not to plant Desert Quandongs near poisonous plants as toxins may be transferred.

The parasite-host relationships need to be further investigated. In many cases, Desert Quandong plants will eventually kill their hosts. Currently *Acacia calamifolia*, *A. hemiteles*, and *A. argyophylla* are believed to be the best Acacia species for hosting Desert Quandong. More research needs to be conducted to determine the best hosts from other genera.

Nutrition:

Quandong kernels are rich in oil and contain the rare fatty acid, Santalbic acid.

Kernels (per 100g):

Energy	698kcal
Water	2.4g
Protein	16g
Fat	63.1g
Carbohydrates	17.1g

Source: An assessment of the composition and nutrient content of an Australian Aboriginal hunter-gatherer diet

Flesh (per 100g):

Energy, including dietary fiber	206kJ
Moisture	71.9g
Protein	2.5g
Nitrogen	0.4g
Fat	0g
Ash	2g
Dietary fiber	4.2g
Fructose	4.6g
Glucose	2.3g
Sucrose	1.2g
Total sugars	8.1g
Starch	0g
Available carbohydrate, without sugar alcohols	8.1g
Available carbohydrate, with sugar alcohols	8.1g

Calcium	7mg
Magnesium	28mg
Phosphorus	11mg
Potassium	747mg
Sodium	116mg
Thiamin (B1)	0.04mg
Niacin Equivalents	0.42mg
Folate, natural	191mcg
Total folates	191mcg
Dietary folate equivalents	191mcg
Alpha carotene	22mcg
Beta carotene	45mcg
Cryptoxanthin	12mcg
Beta carotene equivalents	62mcg
Retinol	0ug
Retinol equivalents	10ug
Vitamin C	20mg

Source: Food Standards Australia New Zealand

Related Species:

Santalum spicatum, also known as Sandalwood, is another parasitic species native to Australia, with the same climatic requirements as *S. acuminatum*. Sandalwood was nearly decimated at the beginning of the 20th century due to its aromatic timber, which contains a sought-after essential oil. The wood is also very useful as a fuel, and the seeds were traditionally eaten by the Aborigines.

Mongongo

Scientific name: *Schinziophyton rautanenii*
Family: Euphorbiaceae
Synonyms: *Ricinodendron rautanenii*; Mongongo Nut, Mongongo Fruit, Manketti Nut, Featherweight Tree
Zone: 9b

Figure 77: *S. rautanenii*. Matetsi Safari Area, Zimbabwe
Credit: Bart Wursten

Overview:

The Mongongo is a dioecious, winter-deciduous tree growing up to a maximum height of 20m (66ft), but usually only to 15m (49ft) tall with an equal spread. The tree develops a straight trunk of up to 1m (3.2ft) in diameter, and has beautiful, compound leaves. The flowers are whitish or yellowish, and give rise to an economically important, edible fruit and seed. From seed, the trees take 15-25 years to start producing fruit, and usually live for more than 100 years.

Habitat and Requirements:

Mongongo is native to Southern Africa, primarily in the arid and semi-arid, summer rainfall regions of southwest Zambia, northern Botswana, and northern Namibia, where it receives between 150-1000mm of average annual rainfall. Light frosts are not uncommon in the winter, and the tree is adaptable to colder areas such as the deserts of Israel. In the wild, it is often found growing with Monkey Orange (*Strychnos cocculoides*). Although the trees perform best during years of average rainfall, they can also survive droughts that last for years.

The trees grow mostly in areas with deep sand, and require well-drained soil. Despite this fact, they do not appear to grow in very rocky soil. They often grow well in soils that are 90% nutrient-poor sand. The trees grow in upland locations on hills, as well as near permanent or ephemeral watercourses. The trees do not tolerate waterlogging or clay soil, and are not very salt-tolerant.

Figure 78: *S. rautanenii* foliage. Credit: N. Juergens

Uses:

Fruit and Seeds:

Under ideal conditions, trees can yield from 830-1000kg of fruits per hectare per year. Most of the fruits fall in April-May (October-November in Northern Hemisphere). The fruits fall off the tree when unripe and ripen a few days later. Provided there are no rains, the fruits may last on the ground for up to eight months. The flesh of the fruit tastes similar to dates, and they are often dried and stored for later use in much the same way as dates. The fruits taste great when consumed raw, but they are usually cooked to help separate the flesh from the seeds. The skin is removed before consuming the flesh.

The fruits usually contain one or two seeds. The seeds have a thick-shell, but contain a delicious kernel. The kernels contain high amounts of fat and protein, and can be eaten raw or cooked. They taste similar to cashews when lightly roasted. The edible oil that is extracted from the seeds deteriorates quickly, but can be used for cooking and also for soap-making.

The seeds as well as the fruit's flesh can also be used as fodder for livestock, and is an esteemed food of elephants.

Wood:
The wood is very lightweight, yet strong, and can be used as a substitute for Balsa wood.

Miscellaneous uses:
The tree can be planted as a windbreak, dune stabilizer, and shade tree, since it forms a dense canopy. In some areas, truncheons are also planted close together in order to form a living fence.

Propagation:

Ethylene gas or soaking in ethephon helps break seed dormancy, and causes many seeds to germinate in as little as one week. The seed coat can also be nicked and the seeds soaked in gibberellic acid for 48 hours prior to sowing. For the first two years of their life, the seedlings grow best in about 50% shade. During their first two years, their growth is slower in heavy shade, and even slower in full sun.

Grafting is unheard of, but probably feasible. The plants are easily propagated from truncheons, which are essentially small branches that are removed and placed directly in the ground.

Potential Drawbacks:

The wood is subject to discoloration by staining fungi, and should be dried immediately after being milled.

Nutrition:

Fruit pulp (per 100g):

Moisture	8.4 - 16.6%
Crude Protein	6.8 – 9.4%
Fat	0.5 – 1.6%
Carbohydrates	65.4 – 76.9%
Sucrose	29.8%
Glucose	0.2%
Fructose	0.5%
Fiber	2.5 – 3.0%
Ash	4.6 – 5.6%

Calcium	85 – 104mg
Magnesium	214 – 266mg
Sodium	1.9 – 2.4mg
Potassium	2145 – 2666mg
Copper	1.3 – 1.6mg
Zinc	1.7 – 1.8mg
Phosphorus	62 – 74.3mg
Iron	1.0 – 4.3mg
Vitamin B1	0.42 – 0.49mg
Vitamin B2	0.13 – 0.21mg
Vitamin C	14.7 – 27mg
Energy	1310 – 1424KJ

Kernel (per 100g):

Moisture	4.1 – 6.5%
Protein	22.0 – 28.8%
Fat	45.0 – 59.6%
Carbohydrates	4.6 – 5%
Fiber	2.5 – 2.7%
Ash	3.7 – 4.1%

Calcium	193 – 225mg
Magnesium	493 – 527mg
Sodium	3.10 – 3.35mg
Potassium	633 – 674mg
Copper	2.52 – 2.82mg
Zinc	3.54 – 4.09mg
Phosphorus	845 – 869mg
Iron	3.42 – 3.7mg
Vitamin B1	0.26 – 0.31mg
Vitamin B2	0.21 – 0.22mg
Vitamin C	2.2mg
Alpha-tocopherol	29mg
Gamma-tocopherol	536mg
Energy	2680 – 2740KJ

Amino acids and fatty acids of kernel:

Essential amino acids (per 100g):

Arginine	3.5g
Histidine	0.7g
Isoleucine	0.7g
Leucine	1.4g
Lysine	0.7g
Methionine	0.4g
Phenylalanine	1.3g
Threonine	1g
Valine	1.8g

Complementary amino acids (per 100g):

Cystine	0.1g
Tyrosine	0.5g

Other amino acids (per 100g):

Aspartic acid	2.4g
Glutamic acid	4.2g
Alanine	1g
Glycine	1.2g
Proline	1.2g
Serine	1.3g

Fatty acids (percent of fatty acids):

Palmitic	9.5%
Stearic	7.6%
Oleic	17.7%
Linoleic	42.9%
Elacostearic	21.7%
Arachidic	0.6%

Source: *Non-timber Uses of Selected Arid Zone Trees and Shrubs in Africa*

Related Species:

Ricinodendron heudelotii is a similar species, which is native to the rainforest regions of Africa.

Marula

Scientific name: *Sclerocarya birrea subsp. caffra*
Family: Anacardiaceae
Synonyms: Southern Marula, Food of Kings, Morula, Elephant's Plum, Drunkard's Plum
Zone: 9b

Overview:

Marula is a dioecious, deciduous tree that can grow to a maximum height of 18m (59ft), but is usually found at half this height. Marula develops a thick, succulent-like taproot, which helps the tree through prolonged droughts. The tree loses its leaves during the winter months, and flowers from September to November (March-May in the Northern Hemisphere). The tree is usually dioecious, but occasionally the trees are hermaphrodites. The hermaphroditic trees usually produce more male flowers than female flowers. From January to March (July-September in the Northern Hemisphere) flowers give rise to edible fruits, which contain an edible nut that has a high protein and fat content. From seed, the trees usually start to bear fruit when they are 6-10 years old. Individual mature trees produce anywhere from 0.2–4.5 tons of fruit in a single season, making Marula a valuable food resource, which has been esteemed since at least 10,000-9,000 B.C.

Figure 79: One-year old *S. birrea ssp. caffra.* Santa Barbara, California
Credit: Mark Viergutz

Habitat and Requirements:

Marula is native to parts of Botswana, Mozambique, Zimbabwe, Namibia, northern South Africa and Swaziland. Across its native range, Marula experiences an average annual rainfall of 200-1370mm. The trees receive the majority of their rainfall during the summer months, and usually experience rainless winters.

In trial plantings in the Negev Desert of Israel, young plants (about 3-5 years old) suffered extensive limb damage when temperatures dropped to -7°C (19.4°F), but nevertheless, resumed growth in the spring. Marula tolerates soil salinity levels of up to about 4 deciSiemens per meter (dS/m). Marula prefers to grow in sandy loam, but also grows well on other deep, well-drained soils such as sand and sandy clay. The trees grow well in areas that have a soil pH of 5-8.

Figure 80: *S. birrea ssp. caffra* fruit. Credit: Ton Rulkens

Uses:

Fruit and Seeds:

For maximum productivity, one male tree should be planted for every five female trees. Fruits weighing over 200g (half a pound) have been developed under cultivation. The fruits ripen in about five days after they fall from the tree. The fruits are usually about the size of a plum and have a sweet, tart taste attributed mostly to their high Vitamin C content. The fruits often spontaneously ferment, which contributes to their employment in a number of alcoholic beverages including Marula Wine/Cider and a liqueur known as Amarula, which is sold in many countries throughout the world. The fruits can be eaten fresh, but are most commonly juiced. The juice from the fruits is also boiled down to a thick, black syrup, which is used as a sweetener.

The flesh of the fruit clasps tightly to a single, central seed similar to how the flesh of a mango grasps its own seed. The hard, woody seed of the Marula fruit contains 1-3 kernels that have a taste similar to that of Macadamia nuts. Due to their high fat content, protein, oil, and calories, the Marula kernels are often more sought-after than the flesh surrounding them. The Kernels yield edible oil that allows individual kernels to be burnt like a candle. Nutritionally, the oil is similar to olive oil, but is considered to be about ten times more stable, and can even be used to preserve meat. The oil is also said to help cure skin ailments, contributing to its extraction for skin-care products and other cosmetics such as anti-aging formulas.

Due to their hard, thick seed coat, the whole seeds can be kept for long periods of time without deteriorating. As such, they are a valuable staple crop. One ton of seeds yields approximately 40kg of oil and 40kg of edible protein.

The fruits are also enjoyed by many wild animals, and can also be used as fodder for livestock. The seedmeal, a byproduct of oil extraction, is often used for this purpose.

Flowers:

The flowers produce lots of nectar, and are favored by bees.

Leaves:

The foliage as well as the fruit and bark are used as fodder for livestock.

Wood:

The wood is easy to work, rather light, yet durable and strong, and does not splinter. Therefore, the wood is widely used by carvers to make various utensils as well as furniture.

Bark:

The bark can be used to treat fever, boils and diarrhea. Inner bark is very fibrous and is used to make a high quality rope.

Miscellaneous uses:

The trees can be coppiced and pollarded, and will often root sucker in addition to resprouting from the trunk. Since Marula roots easily from truncheons, it can be used as a living fence by planting many truncheons close together. The tree also hosts the edible mopane caterpillar, which provides a famine food.

Propagation:

Seeds have a dormancy period, and will not germinate when they are very fresh. Many seeds will germinate up to one year after they are planted. Seeds can be soaked overnight in water, but this is not necessary. The seeds have a very thick, hard, and woody seed coat, which renders nicking futile. Individual seeds can be placed in plastic bags filled with damp peat moss, and then transplanted after they germinate. Otherwise, the seeds are best planted directly, or in individual, deep-rooted pots. Seeds should be sown from spring to early summer, and most seeds will germinate within a couple months.

The seedlings can be kept in deep-rooted containers, but should generally be planted out within two years due to their large, succulent taproots, which develop much more quickly than the above ground growth.

The plants can also be propagated from cuttings and by separating root suckers. Marula trees are easily propagated from truncheons up to 2 meters long with a girth of up to 150mm. Grafting is said to be up to 95% successful when using tip cuttings 5-10cm in diameter as scions. Therefore, elite trees can be easily selected and propagated. Female scionwood can be grafted on to male trees in order to obtain more productive trees per unit of area. Vegetative propagation, including grafting should be done just as dormancy ends in the spring.

Potential Drawbacks:

Since the flesh adheres tightly to the seed, it can be difficult to separate. Also, the kernels are difficult to extract because of their thick, hard seed coat. A macadamia nutcracker, vice, or other heavy-duty decorticator should be used to facilitate kernel extraction.

Nutrition:

Marula flesh can contain up to 180mg of Vitamin C per 100g, which is about four times the amount of an orange. The kernel and kernel oil are also very nutritious. The kernels typically contain 700-750 calories per 100g.

Fruit flesh (per 100g):

Moisture	91.7g
Ash	0.2g
Protein	0.5g
Fat	0.1g
Fiber	0.5g
Carbohydrates	7.0g
Vitamin C	67.9mg
Calcium	6.2mg
Magnesium	10.5mg
Iron	0.1mg
Sodium	Trace
Potassium	54.8mg
Copper	0.04mg
Phosphorus	8.7mg
Thiamin	0.03mg
Riboflavin	0.05mg
Nicotinic Acid	0.25mg

Seeds (per 100g):

Moisture	4.0g
Ash	4.2g
Protein	30.9g
Fat	57.0g
Fiber	2.4g
Carbohydrates	1.5g
Calcium	106mg
Magnesium	467mg
Iron	0.42mg
Sodium	338mg
Potassium	677mg
Copper	1.99mg
Phosphorus	836mg
Thiamin	0.04mg
Riboflavin	0.12mg
Nicotinic Acid	0.71mg

Source: SEPASAL Database

Fatty acid content of kernel oil (per 100g):

Lauric Acid	Trace
Palmitic Acid	10.7g
Stearic Acid	6.9g
Oleic Acid	72.0g
Linoleic Acid	8.8g
Linolenic Acid	0.6g
Gadoleic Acid	0.7g
Behemic Acid	0.1g
Erucic Acid	Trace
Total Saturated	17.7g
Total Monounsaturated	72.8g
Total Polyunsaturated	9.4g

Source: The Journal of Food Technology in Africa

Related Species:

There are three subspecies of Marula, including *Sclerocarya birrea ssp. caffra*. Of the other two subspecies, *Sclerocarya birrea ssp. birrea* is native to West Africa into Tanzania, while *ssp. multifoliata* is native to central Tanzania. These subspecies are not as important as *ssp. caffra.*

Monkey Orange

Scientific name: *Strychnos cocculoides*
Family: Loganiaceae
Synonyms: Bush Orange, Monkey Apple, Corky-Bark Monkey Orange, Moruta
Zone: 9b

Overview:

The Monkey Orange is a drought-deciduous tree that can attain heights of up to 8m (26ft) but usually maxes out at about 5m (16ft). The lateral, canopy spread of the tree often measures wider than the height of the tree. In arid climates the tree grows deep roots and is able to withstand prolonged droughts by shedding its leaves and remaining deciduous until rain returns. The tree is covered with spines longer than 1cm, and the plants frequently sucker. From seed, trees begin to bear fruit after 4-6 years. Fruits ripen in October to December (April-June in the Northern Hemisphere). These fruits contain a sweet pulp, which is highly sought-after.

Habitat and Requirements:

The Monkey Orange is found throughout southern Africa and into equatorial east Africa. The plant is found from northern South Africa, eastern Botswana, Zimbabwe, northern Namibia, Angola, Mozambique, Democratic Republic of the Congo, Zambia, Tanzania and into southern Kenya. Throughout this range, the plant grows in savannas as well as woodlands, where it often experiences extreme heat.

Most sources site Monkey Orange as growing in areas with 600-1500mm of average annual rainfall, however, evidence suggests that the plant can grow in areas of east Botswana that receive only 200mm of mean rainfall per year. In locations that receive such low rainfall, the tree is most likely found where there is a high water table. The plant's optimal rainfall preference is between 700-1200mm, and there is

Figure 81: *S. cocculoides*. Katanga, Democratic Republic of the Congo
Credit: Ingrid Parmentier

a correlation with a lower percentage of fruit pulp in fruits from areas with less rainfall. Monkey Orange grows in summer rainfall regions that experience drought during the winter months.

Although Monkey Orange is usually found growing on sandy soil, the tree has a wide tolerance for soil types, tolerating clay as well as rocky soil. The tree is known to tolerate a pH range from 4-8, but grows optimally in pH 5.5-7.5 soils. The plant often grows as a pioneer in areas that were once used for agriculture.

Uses:

Fruit and Seeds:

Individual mature trees produce anywhere from 300-700 fruits per year with a total fruit weight of 40-100kg. The fruits are 6-12cm in diameter with a 2-5mm thick shell, and contain 10-100 seeds. The pulp is the edible part of the fruit, and usually comprises 40-50% of the entire fruit (seeds and shell included). When ripe, the fruits fall from the trees, and can be hit lightly against a rock in order to break the shell. The seeds are spat out, or the pulp can be pressed through a screen. The pulp has a sweet flavor, and is made into alcoholic beverages such as wine or liqueur. The fruits are also juiced, made into jam, or dried for later use. Additionally, the pulp is often mixed with honey or sugar as a folk remedy for coughing.

Fruits should be eaten only when fully ripe, because unripe fruits contain alkaloids as well as high amounts of saponins. Fruits that are knocked off the tree before fully ripening will ripen in a few days at room temperature. The saponins contribute to the unripe fruits' usage as a soap substitute. Also, unripe fruits are sometimes consumed to induce vomiting. The hollowed shells of the fruits are used as resonance chambers for musical instruments.

Leaves:

The leaves can be cooked in water to make a porridge, which is applied externally to wounds to foster healing and prevent infections.

Wood:

The wood is relatively light, and is used to make posts, tool handles, and other utensils.

Bark:

The bark is used as a cure for stomach aches by cooking the middle portion of bark in water and drinking the resultant liquid.

Roots:

The roots are chewed to combat gonorrhea and other sexually transmitted infections, as well as to treat eczema.

Miscellaneous uses:

The plants can be coppiced, and grow back very well, readily producing shoots.

Propagation:

The seeds require no special pre-treatment, but can be soaked in warm water for 1-2 days before sowing at a depth of 2-3cm. The seeds take from 1-4 months to germinate, and germination rates of 75-80% can be expected for fresh seeds. Viability is lost when the seeds are stored for long periods of time, so they should be used sooner rather than later. Optimal storage is attained at temperatures slightly above freezing. Once seedlings are large enough, they should be transplanted to deep-rooted containers.

The trees should be kept in partial shade (50%) for the first year or two of their lives. Shade is particularly useful in the nursery as it discourages root development, something that may quickly lead to pot-bound plants.

Although the trees do not root from cuttings or from separated suckers, they can be grafted rather successfully. The preferred method for grafting is the wedge method; however budding techniques are also effective.

Potential Drawbacks:

The seeds contain small quantities of strychnine, and should not be eaten. For children, the seeds may be difficult to avoid. Although *Strychnos cocculoides* produces the sweetest fruit pulp of all the *Strychnos* species, the seeds of *S. pungens* and *S. innocua* contain no alkaloids, and are safer if accidentally consumed.

Nutrition:

Mineral content of fruit pulp (mg/g):

Phosphorus	20.2mg
Calcium	9.41mg
Magnesium	26.9mg
Iron	0.18mg
Potassium	188mg

Source: Southampton Centre for Underutilised Crops

Related Species:

Unlike *Strychnos cocculoides*, *S. madagascariensis*, *S. spinosa*, *S. pungens* and *S. innocua* do not produce strychnine-containing seeds, and are therefore safer to consume. The drawback, however, is that none of these fruits have pulp as sweet as the fruit pulp of *S. cocculoides*. The closest rival is perhaps *S. spinosa*, whose pulp has a sour-sweet flavor. The wood of this species is harder, and better suited for construction.

Jujube

Scientific name: *Ziziphus jujuba*
Family: Rhamnaceae
Synonyms: Chinese Date, Chinese Jujube, Common Jujube, Sedra, Korean Date Tree
Zone: 4a

Figure 82: *Z. jujuba*. Cerrilos, New Mexico
Credit: Amanita Thorp. SunStar Herbs

Overview:

The Jujube is a deciduous tree, which can attain a height of 12 meters (39ft), but usually only grows up to 9m (30ft). The drooping canopy has a diameter of about 4-5m (13-16ft). The branches grow in a zigzag pattern, and are adorned with sharp spines. The oval leaves add ornamental value to the autumn landscape when they turn bright yellow. Jujube trees bloom over the course of a few months from late spring into summer, and produce delicious, edible fruits that turn from green to red. Most trees do not require cross-pollination for fruit set, but in such cases the seeds that are produced are not viable. Plants are fast-growing and fruit after 3-4 years when grown from seed.

Habitat and Requirements:

Jujube is native to the Loess Plateau of central China. This is a semi-arid region that receives 200-750mm of average annual rainfall, mostly in the form of summer monsoons. The region has a continental climate with cold, harsh winters, and very hot summers. Jujube is widely adaptable; growing in areas with mild winters as well as areas with harsh winters, and grows well in areas that experience extreme heat during the summer months. As an adaptation to regions with unpredictable cold spells, the plant displays delayed budding in order to avoid damage from late, spring frosts.

Even at a young age, the plants require full sun. They grow best in areas of deep, sandy loam soils, and generally prefer substrates with good drainage. The plants grow much more slowly in heavy clay soil. Jujube trees are salt-tolerant, and are able to grow in areas with salinity levels of up to 10-15dS/m (deciSiemens per meter). Jujubes grow adequately with a soil pH of 4.5-8.4.

Uses:

Fruit and Seeds:

Lack of heat in summer as opposed to insufficient winter chill is the limiting factor for fruit set. Where summers are warm enough, mature trees produce up to 9.5kg of refreshing, cherry or plum-size, red drupes per year. The fruits are often preferred when they are just starting to turn red. At this stage they possess a taste similar to an apple. When allowed to dry on the trees, the fruits are not as flavorful, but have the advantage of acquiring an indefinite shelf life. The dried fruits can be stored whole or ground into a powder. The seeds have anxiolytic and sedative effects.

Leaves:

The leaves can be cooked and eaten as a famine food when nothing else is available.

Wood:

The wood is very hard and tough, and can be used to make agricultural tools. It is a very high quality fuelwood, and can be made into charcoal.

Propagation:

Fresh seeds should be sown in the spring after soaking in water for 12-24 hours. Older seeds appear to go dormant, and should be subjected to a three-month-long warm stratification followed by a three-month-long cold stratification. Most seeds germinate within a couple months; however, some seeds may take up to one year before sprouting. The plants are commonly grafted and budded in order to propagate reliable cultivars. Cuttings of the current season's growth can be taken in the winter, but they do not appear to root very successfully. On the other hand, root suckers may be removed and planted elsewhere during the winter. These separated root suckers are much more likely to develop into a mature plant. The plants tolerate container culture; however this is not ideal, since the plants develop deep taproots.

Potential Drawbacks:

Although the plants can be heavily pruned and coppiced, doing so results in vigorous root suckering. Even on spineless, grafted plants, the rootstock is often heavily armed with spines, which means there is the potential for ongoing removal of spiny suckers. The suckers are removed by mowing or hoeing.

Nutrition:

Dried Fruits (per 100g):

Water	19.70g
Energy	287kcal
Protein	3.70g
Total Lipids (Fat)	1.10g
Carbohydrates, by difference	73.60g
Calcium	79mg
Iron	1.80mg
Magnesium	37mg
Phosphorus	100mg
Potassium	531mg
Sodium	9mg
Zinc	0.19mg

Vitamin C, total ascorbic acid	13.0mg
Thiamin	0.210mg
Riboflavin	0.360mg
Niacin	0.500mg

Source: USDA National Nutrient Database

Christ's Thorn Jujube

Scientific name: *Ziziphus spina-christi*
Family: Rhamnaceae
Synonyms: Christ's Thorn, Syrian Christ Thorn, Crown of Thorns Tree
Zone: 9b

Overview:

Christ's Thorn Jujube is a semi-deciduous tree (evergreen in warm winter areas) that is believed by many to be the tree from which Jesus' crown of thorns was made. Indeed, it is a thorny tree, which, under ideal conditions, can grow up to 20m (66ft) tall. However, it usually only grows up to 10m (33ft). Often the tree forms a thorny, impenetrable thicket. The tree has whitish branches, and a drooping form similar to that of the common Jujube. The tree produces copious amounts of edible, reddish-orange fruits with a diameter of about 2cm.

Habitat and Requirements:

Christ's Thorn Jujube is native to the Sahara Desert and the Sahel, a strip of land bordering the southern end of the Sahara Desert. The tree is believed to be native to most of arid, northern Africa – from Mauritania to the Red Sea, but nonetheless, the tree has also naturalized in parts of southern Africa, the Arabian Peninsula, and the Mediterranean Basin. This naturalized distribution range shows that Christ's Thorn Jujube can grow in areas of winter rainfall, summer rainfall, or both. It grows well in areas that receive 100-500mm of average annual rainfall, and tolerates prolonged periods of extreme heat. The tree has an extremely deep taproot, and prefers to grow wherever there is groundwater.

The tree tolerates a similar soil pH range as the common Jujube, but is slightly more salt-tolerant, being able to grow in areas where soil salinity measures at least 15.2dS/m (deciSiemens per meter). The tree prefers well-drained, deep, alluvial soil, but also grows well in heavy soil, and is able to tolerate waterlogging for up to two months.

Figure 83: *Z. spina-christi* fruit. Credit: Bernadette Simpson

Uses:

Fruit and Seeds:

The fruits are sweet and contain about 15% sugar. The seeds are also edible, but their hard shell should be removed, and the seeds should be roasted.

Flowers:

The flowers are esteemed in apiculture.

Leaves:

The foliage can be used as fodder for livestock. Humans can also cook the leaves and eat them as a famine food.

Wood:

The wood is dense and can be used for fuel or made into an excellent charcoal. Since the wood is termite-resistant, it is often used for construction and to make handles for agricultural implements. Despite their slow growth rate, the trees coppice very well, and can thus be grown sustainably for their timber.

Miscellaneous uses:

The trees can be planted densely to form a living fence, or individual branches can be used as fencing material. Due to its vigorous root structure, with a deep taproot and extensive lateral roots, the tree is a good candidate for stabilizing sand dunes and preventing erosion. The tree is also a good phosphorus accumulator, which can be pruned, and used as mulch for intercropped plantings.

Propagation:

The hard, woody shells should be removed, and the seeds should be soaked in water for two days. The seeds can also be soaked in sulfuric acid for a couple hours, and then cold stratified before being sown. Due to their taproot, they should be sown directly, or in deep-rooted containers. After half a year in the nursery the plants should be planted out.

The trees can also be propagated via root suckers and woody cuttings. Christ's Thorn Jujube trees are often used as rootstock for the Indian Jujube (*Ziziphus mauritiana*), and can thus be budded and grafted successfully onto each other.

Potential Drawbacks:

Like the common Jujube, the plants send out spiny suckers, which may be undesirable. Coppicing and heavy pruning tends to increase the frequency of suckers. This species of Jujube has a slower growth rate than the common Jujube.

Nutrition:

Various Plant Parts (per 100g):

	Fruits:	Seeds:	Leaves:
Carbohydrates	63.8%	21.8%	25.8%
Protein	3.1%	29.6%	6.8%
Fat	2.2%	3.9%	3.3%
Calcium	177mg	154mg	1270mg
Iron	0.6mg	4.4mg	7.2mg
Phosphorus	135mg	1090mg	85.4mg
Sulphur	94.5mg	1180mg	195mg
Potassium	1910mg	1130mg	673mg
Magnesium	56.3mg	301mg	169mg
Zinc	0.8mg	9.2mg	1.5mg
Copper	0.6mg	1.6mg	0.3mg
Manganese	0.4mg	3.5mg	3.5mg
Sodium	9mg	14.1mg	22mg

Source: Eden Foundation

Related Species:

Ziziphus mucronata is found commonly across the dry regions of southern Africa. There it is valued for its edible fruits and seeds, which are roasted, ground and used as a coffee substitute. *Ziziphus lotus* is a shrub native to the Mediterranean region of Tunisia, which receives approximately 150-200mm of average annual rainfall. *Ziziphus nummularia* is a shrub native to the Thar Desert of Pakistan and western India, sharing a similar distribution range as Moth Bean (*Vigna aconitifolia*). All of these species of *Ziziphus* produce edible fruits.

Ziziphus mauritiana is very similar to *Z. spina-christi*, and is often grafted onto *Z. spina-christi* rootstock. *Z. mauritiana* has received much more research attention, however, and is now a semi-domesticated crop for the arid and semi-arid tropics and subtropics. It is not quite as drought-tolerant as *Z. spina-christi*.

Categorized Plant Reference Lists

The following lists refer to the plants showcased in parts 1-4 of this book. The plants are grouped into categories based on various characteristics, and can be used as quick references for situations where a specific trait is or is not desired.

Minimum Rainfall Requirements

The following rainfall requirements are approximations intended to give a general idea. For more detailed information, please consult the main parts of the book.

Plants that require 100mm or less of annual rainfall:

Plants that require a minimum of 101-199mm of annual rainfall:

Plants that require a minimum of 200-299mm of annual rainfall:

300-399mm of annual rainfall:

400-499mm:

500+mm:

Annual Plants or Plants Grown as Annuals

Plants That Can Be Coppiced/Pollarded

Deciduous Species

Dioecious Species (plants with male and female flowers on separate plants)

Nitrogen-Fixing Plants

Salt-Tolerant Species

Shade-Tolerant Species

Plants With Spines/Thorns

Noteworthy Timber Species

Potential Crops Deserving Further Investigation

Acacia senegal – A deciduous tree native to Sub-Saharan Africa. The tree is the primary source for gum arabic. The seeds are edible.

Agave tequilana – Native to western Mexico. This plant is commonly used for the production of tequila. A crude molasses product can also be obtained from concentrated plant nectar. Recently, the plant has been used to produce highly refined fructose syrup similar to high fructose corn syrup. This syrup is often marketed as "agave nectar" or "agave sweetener."

Amelanchier alnifolia – A deciduous shrub native to central Canada, the central-northern U.S. and the pacific northwest, growing as far north as southern Alaska. The fruits are similar in appearance, flavor and nutritional content to blueberries.

Anacampseros australiana – A small succulent native to arid and semi-arid parts of Australia. The tuberous roots can be eaten.

Annona senegalensis – Native to tropical Africa. The fruits contain sweet pulp, and are similar to the widely cultivated Cherimoya (*Annona cherimola*).

Ariocarpus agavoides – A small succulent species with edible roots native to the Chihuahuan Desert of Mexico.

Atriplex halimus – An evergreen shrub native to northern Africa and the Mediterranean Basin. The leaves can be used as a vegetable crop, and have a salty flavor.

Azadirachta indica – A tree native to northern Myanmar and eastern India. The tree is best-known for producing neem-oil, which has insecticidal properties and a myriad of other uses.

Azanza garckeana – A native of semi-arid southern and eastern Africa. This semi-deciduous tree grows to 3-15m in height and grows with as little as 250mm of rainfall per year. The tree tolerates clay soil as well as light frost. The sweet, nutritious fruit flesh is chewed raw like chewing gum, or boiled into a syrup and used as a sweetener for other foods. The leaves can also be eaten cooked.

Bauhinia petersiana subsp. macrantha – A large shrub native to northwestern South Africa, Namibia, Botswana, Zimbabwe, Zambia and Angola. The shrub produces very showy, white flowers, and edible seeds that can be used as a pulse.

Beilschmiedia berteroana – An evergreen tree native to central – southern Chile. The fruits are said to be edible.

Berchemia discolor – A tree native to semi-arid regions of southern Africa. The tree grows well in areas that receive as little as 250mm of annual rainfall. The tree is frost tender, and dislikes cold wind. The tree produces edible fruits and seeds. The fruits taste like dates, and the seeds taste like walnuts.

Bombax ceiba var. leiocarpum – Native to the dry forests of northern Australia. The fleshy taproots of younger plants are eaten roasted. The flowers and young leaves are used as a spice.

Borassus aethiopum – A tall fan palm native to the savannas of Sub-Saharan Africa. Fruit pulp can be consumed raw or cooked. Immature seeds contain a jelly-like, sweet substance. The seedlings can be used as a vegetable.

Borassus flabellifer – Native to India and northern Myanmar, as well as the surrounding region. Every part of the palm is used for something. The fruit and seeds are edible. Most commonly, however, the inflorescence is tapped for its sweet sap, which can be drunk without any processing. Otherwise the sap is usually converted into sugar, or palm wine.

Boscia albitrunca – An extremely deep-rooted small tree that is native to dry areas of southern Africa. Live tree roots in the Kalahari have been found at a depth of 68m (223ft). The roots and fruits are edible.

Boscia senegalensis – A small evergreen tree native to the Sahel. The fruits are edible. The seeds require leaching with water in order to be rendered edible.

Brassica carinata – Native to semi-arid east Africa. The plant is also adaptable to temperate regions. The leaves supposedly taste better than Kale, and are rich in vitamin C, vitamin K, beta-carotene, calcium and anti-oxidants. Seeds produce edible oil that is similar to Canola oil.

Calandrinia balonensis and *C. polyandra* – Native to arid parts of Australia. The leaves, roots and seeds of these small succulents can be eaten raw, but should be cooked due to the presence of oxalic acid.

Caragana arborescens – Large, deciduous, nitrogen-fixing shrub native to Siberia and Manchuria. The mature seeds can be boiled with one change of water and eaten.

Cercidium microphyllum and *floridum* – These small nitrogen-fixing trees are native to the Sonoran Desert of North America. Their seeds and flowers are edible. The trees provide a dappled shade habitat useful for establishing other desert plants.

Citrullus colocynthis – An herbaceous, perennial, groundcover cucurbit native to the Sahara Desert. The fruits look like bright-yellow balls, and can roll across the desert during windstorms. The fruit pulp is poisonous, yet, along with the foliage, contains anti-cancer and anti-tumor compounds. The seeds, on the other hand, are edible, calorie-rich and contain high amounts of protein and oil.

Colophospermum mopane – Native to dry regions of central and southern Africa. The fruits of this large tree can be used as a famine food. The tree hosts the Mopane Caterpillar, which is commonly eaten as a protein-rich food. Therefore, this tree is particularly valuable to non-entomophobic cultures.

Cucurbita foetidissima – A perennial cucurbit with triangular leaves that is native to the southwestern United States and northern Mexico. The young, immature fruits can be cooked and eaten as a vegetable. The mature seeds are edible, containing high amounts of oil and protein. The large roots are a source of starch.

Cynara cardunculus – This native of the western Mediterranean region is the wild relative of the globe artichoke. The flower buds can be consumed in the same way as those of the globe artichoke. The roots and stems can be eaten cooked. The young leaves can be eaten raw or cooked.

Dialium schlechteri – A deciduous tree of the pea family growing up to 15m in height. The tree is native to the Maputaland sand forests in northern Kwa-Zulu Natal, South Africa. The fruits are edible, and reportedly taste like sherbet.

Dioscorea remotiflora – A climbing vine native to the subtropical, dry, deciduous forests of western Mexico. The large tubers are harvested in late summer – early fall and can be consumed raw, but are usually eaten cooked like yams.

Diospyros mespiliformis – A large evergreen tree native to semi-arid regions of southern Africa. The tree is valued for its edible fruits, and for its wood, which is a high-quality source of fuel and timber.

Echinocereus stramineus – A species of cactus native to the southwestern United States. The fruits taste similar to strawberries.

Echinocereus triglochidiatus – A species of cactus native to the southwestern United States. The fruits are sweet.

Escontria chiotilla – Native to semi-arid portions of southern Mexico.

This tree cactus produces many small fruits that taste similar to blueberries.

Espostoa lanata – A species of columnar cactus native to the deserts of northern Peru and southern Ecuador. The hairs that the cactus produces are used as pillow stuffing. The reddish-purple fruits are edible and sweet.

Eugenia zeyheri – An evergreen shrub native to semi-arid parts of southern Africa. The plant produces reddish-purple, edible fruits that are the size of cherries.

Ferocactus macrodiscus: A small cactus from very dry areas of central and southern Mexico. The flesh of the entire cactus is commonly combined with sugar and eaten.

Ficus sycomorus – This relative of the common fig is a large, semi-deciduous tree found growing in savanna regions of Africa. Like the common fig, the trees produce edible fruit.

Flacourtia indica – The exact origin of this tree is unkown, however, it is found growing in the dry forests of Africa and India. The tree produces edible, tart fruits.

Geoffroea decorticans – This deciduous tree is native to semi-arid regions of South America, with the largest populations found in northern Chile, and northern Argentina. The tree has similar climatic requirements to *Prosopis alba*. *G. decorticans* is a nitrogen-fixing tree and can be inoculated with *Rhizobium tesota*, and the same *Rhizobium* strains used to inoculate Mesquite trees. Its pods have a high sugar content, and its seeds contain high levels of protein and fat.

Grewia bicolor – A large deciduous shrub native to semi-arid tropical Africa. The plant is valued for its juicy, sweet fruits, which can be eaten raw or fermented into an alcoholic beverage.

Harpephyllum caffrum – A large evergreen tree native to southern Africa. The plant is valued primarily for its edible, plum-like fruits.

Hippophae rhamnoides – A deciduous, dioecious, nitrogen-fixing shrub native to semi-desert regions of central Asia. It is a very cold-tolerant shrub and produces acidic berries that are highly nutritious.

Kigelia africana – A drought-deciduous tree native to tropical Africa. The tree produces very large (up to 12kg), sausage-like fruits that are poisonous raw, but are consumed roasted or fermented. The fruit extracts are commonly used for beauty products and can be used to treat eczema and psoriasis.

Lesquerella fendleri – A small, herbaceous, yellow-flowered plant that is native from Oklahoma and Texas to southeastern Arizona. The plant is cultivated as a winter annual, and is cultivated for its high-quality seed oil, which is a replacement for castor oil. The seed meal can be used as animal feed.

Maerua crassifolia – Native to arid parts of northern Africa. The leaves and fruits of this tree can be eaten. The fruits are said to be sweet.

Medemia argun – A very rare, critically endangered fan palm native to the Nubian Desert of Sudan. Probably the most drought-tolerant palm species in the world. The fruits are similar in appearance to plums, and they, along with the seeds, are reportedly edible.

Mimusops angel – A large evergreen tree native to northeastern Somalia. The tree is valued primarily for its edible fruits.

Mimusops laurifolia – Native to northeastern Africa and parts of Yemen. This evergreen tree is valued for its edible fruits, which are similar to an Avocado.

Mimusops zeyheri – This evergreen tree grows throughout semi-arid southern Africa, and can reach heights of 15m with an equal spread. The tree prefers summer rainfall regions. The sweet fruits are rich in Vitamin C.

Moringa peregrina – Native to the region bordering the Red Sea. The tubers of very young trees can be roasted and eaten. The seed oil is valued as a substitute for olive oil, and has a greater resistance to rancidity.

Nannorrhops ritchiana – A clumping fan palm native to the Middle East. Grows well in dry areas with cold winters and very hot summers. The seeds are edible, and the young leaves are edible either raw or cooked.

Neowerdermannia vorwerkii – A small cactus native to the Andes Mountains of Bolivia and northern Argentina. The entire plant is eaten cooked.

Olneya tesota – A small nitrogen-fixing tree native to the Sonoran Desert of North America. The seeds and flowers are edible. The trees provide a dappled shade habitat useful for establishing other desert plants.

Oreocereus pseudofossulatus – A columnar cactus native to high altitudes of the Andes Mountains in central Bolivia. The edible fruits are large and yellow.

Parinari curatellifolia – This tree grows throughout semi-arid, Sub-Saharan Africa. It is a large tree growing up to 20-26m in height with an equal or greater spread. The trees tolerate occasional light frost and prefer well-drained clay-loam soil. The fruits and seeds are edible. The fruit flesh is sweet and high in Vitamin C. The seeds are somewhat similar to almonds, and are rich in oil.

Parthenium argentatum – A small shrub native to the Chihuahuan Desert of North America. The shrub can be used as a source of rubber.

Portulaca oleracea – Native to northern Africa. All plant parts are edible. The leaves are especially a nice vegetable crop, and can be used in salads.

Rhynchosia spp. – Many species are native to arid regions of Africa, where their roots are said to be eaten.

Salvadora persica – An evergreen tree most likely native to the Middle East, and then spread to Africa and India. The leaves can be cooked and eaten as a vegetable. The fruits are sweet and have a slight peppery flavor. The young stems have antibacterial properties, and can be used as a toothbrush substitute.

Sarcozona praecox – A groundcover succulent from dry parts of southern Australia. The fruits are reportedly edible, juicy, and sweet.

Senna petersiana – Native to semi-arid parts of Africa. This small tree is very medicinal, and is often used to combat a variety of diseases and infections. The pulp of the pods is reportedly sweet, and can be eaten raw. The pods are also soaked and boiled before being consumed.

Simmondsia chinensis – A dioecious, evergreen shrub native to the Sonoran and Mojave deserts of North America. This is the plant responsible for Jojoba oil. Jojoba oil has extremely important industrial uses, and is used as a substitute for sperm whale oil.

Stenocereus stellatus – A columnar cactus native from central to southern Mexico. The fruits, seeds, flowers and stems are edible.

Sterculia africana – A caudiciform, deciduous tree native to southern and eastern Africa. The seeds are eaten roasted.

Stomatostemma monteiroae – A climbing succulent with a tuberous root. The plant is native to semi-arid parts of southern Africa. Cooked and dried fruit flesh is used as a vegetarian meat substitute.

Suaeda australis – Native to coastal areas of Australia. *Suaeda spp.* have similar cultural requirements, and are used as leaf vegetables in the same way as *Salicornia spp.*

Tabernaemontana elegans – A small deciduous tree native to semi-arid parts of east and southeast Africa. The fruits contain an edible yellow pulp.

Tetragonia implexicoma – From coastal areas of New Zealand and southern Australia. The leaves of this salt-tolerant, groundcover shrub are used like spinach raw or cooked. The plant also produces small, red berries, which are edible.

Tetragonia tetragonoides – Native to coastal areas of New Zealand. The plant is now semi-domesticated worldwide, and serves as a more drought-tolerant and salt-tolerant substitute for spinach.

Trichocereus validus – A columnar cactus native to the Bolivian Andes. The large fruits are edible and have a sweet flavor.

Uapaca kirkiana – Native to semi-arid regions of eastern to southern Africa. The plum-sized fruits are sweet, high in Vitamin C and highly esteemed. The wood is also valued for fuel and for charcoal making.

Vangueria infausta – A small deciduous tree native to semi-arid parts of eastern and southern Africa. The fruits have a sweet-sour taste and can be eaten raw or cooked. The seeds can be eaten roasted.

Vittelaria paradoxa – A deciduous tree native to semi-arid central Africa. The fruit pulp is edible, and tastes similar to a pear. The plant, however, is mostly valued for its seed kernels, from which shea butter is produced.

Wilcoxia striata – A thin-stemmed, spindly cactus native from central Sonora to southern Baja, Mexico. The bright red fruits contain sweet flesh.

Ximenia americana – A large, deciduous, semi-parasitic shrub that can be found throughout many tropical and subtropical semi-arid regions of the world. The edible fruits are described as sweet and tart. The flower petals are also said to be edible.

Ximenia caffra – A deciduous, hemi-parasitic tree reaching a height of 6m. The tree is native to semi-arid parts of eastern and southern Africa, growing in areas with as little as 250mm of average annual rainfall. The fruits and seed nuts can be eaten. The fruits are extremely high in Vitamin C, and the seeds yield a high-quality oil.

Resources

Inoculants (for nitrogen-fixing plants):

World Federation for Culture Collections
– http://www.wfcc.info/home/ - A vast, worldwide network for finding a broad spectrum of microorganisms, including nitrogen-fixing bacteria.

Plant Probotics
- Tom Wacek (tomwacek@yahoo.com) - A great independent source, located in the U.S.A. with the knowledge to produce his own strains of inoculants that may be difficult to find from other sources. Very helpful.

Becker Underwood
- http://www.beckerunderwood.com/ - A large company with a worldwide distribution. They mostly produce typical, run-of-the-mill inoculants, but also carry various strains for Lupins. The Australian branch carries inoculants for Acacia species.
Becker Underwood Corporate Headquarters
801 Dayton Avenue
Ames, IA 50010 USA
800-232-5907
request@beckerunderwood.com

New Edge Microbials
- http://www.microbials.com.au/ - This Australian based company has a decent selection.
New Edge Microbials Pty Ltd
951 Garland Avenue
Albury NSW

2640
P: (02) 6025 0044
F: (02) 6040 0237
E: newedge@microbials.com.au

Seeds and Plants:

Silverhill Seeds
- http://www.silverhillseeds.co.za/ - An excellent seed vendor that carries most of the African species mentioned in this book.
P.O. Box 53108
Kenilworth
7745 Cape Town,
South Africa
Telephone: +27 21 762-4245/ Facsimile: +27 21 797-6609
E-mail: info@silverhillseeds.co.za

DELEP (Desert Legume Program)
- http://cals.arizona.edu/desertlegumeprogram/seminum.html - Provides small accessions of seeds of many different species of legume plants from around the world.
Seed Bank
Desert Legume Program
The University of Arizona
2120 East Allen Road
Tucson, Arizona 85719
Telephone: (520) 647-2503
E-mail: mjohnson@ag.arizona.edu
Fax: (520) 647-2638

CactusStore.com
- http://www.cactusstore.com/ - Carries seeds and live plants of many species of cacti and succulents.
Phoenix Desert Nursery
3525 E. Southern Ave.
Phoenix, AZ 85040
602-305-9275 (office voice mail)
richarde4@gmail.com

Sacred Succulents
- http://sacredsucculents.com/ - Carries many useful species of cacti and succulents as well as other rare and useful plants from around the world. Sacred Succulents
PO Box 781
Sebastopol, CA 95473
Email: sacredsucculents@hushmail.com

Chileflora
- http://www.chileflora.com/ - A great selection of native Chilean plant seeds.
2 Sur 665
Dpto. 208
Talca, Chile
michail@chileflora.com

B & T World Seeds
- http://b-and-t-world-seeds.com/ - A company that has a very extensive inventory of seeds from around the world.
Route des Marchandes
Paguignan
34210 Aigues-Vives
France
fax ++ 33 (0)4 68 91 30 39
phone ++ 33 (0)4 68 91 29 63
heather@b-and-t-world-seeds.com

Rarepalmseeds.com
- http://www.rarepalmseeds.com/index2.shtml - A large selection of palm tree seeds.
Tobias W. Spanner
Am Schnepfenweg 57
80995 Muenchen
Germany
Phone: (+49) 89.1577902
Fax: (+49) 89.15704284
mail@rarepalmseeds.com

Micologia Forestal & Aplicada
- http://www.micofora.com/ - Sells saplings of useful tree species such as Stone Pine (*Pinus pinea*) that have been infected with truffle fungi.
Rbla. Arnau 6 local D
Vilanova i la Geltrú
08800 Barcelona
Spain
Tel/Fax: +34 93 815 54 55
micologiaforestal@micofora.com

Organizations:

The Seawater Foundation
- http://www.seawaterfoundation.org/ - Focuses on growing salt-tolerant crops using seawater in coastal desert regions.
4500 N. 32nd St
#Suite 203
Phoenix, AZ 85018
Phone: 602.808.4490
Fax: 602.808.4495

The Klein Karoo Sustainable Drylands Permaculture Project
- http://berg-en-dal.co.za/ - One of the most impressive dryland permaculture projects.

Food Plants International
- http://www.foodplantsinternational.com/ - An organization that is building a database of all the edible plant species of the world.
info@foodplantsinternational.com

Desert Harvesters
- http://www.desertharvesters.org/ - An organization based in Tucson, Arizona that teaches how to utilize the native food resources of the Sonoran Desert.
Desert Harvesters
c/o 813 N 9th Ave
Tucson, AZ 85705
hammermill@desertharvesters.org

SEPASAL (Survey of Economic Plants for Arid and Semi-Arid Lands)
- http://www.kew.org/ceb/sepasal/ - An extensive database that provides information on various edible and useful dryland plant species.
SEPASAL, Jodrell Laboratory
Royal Botanic Gardens, Kew
Richmond, Surrey, TW9 3AB
UK
Tel: +44 (0)20 8332 5385
Fax: +44 (0)20 8332 3717
email: sepasal@kew.org

Crops For The Future
- http://www.cropsforthefuture.org/ - An organization that promotes the use of alternative crops worldwide.
Crops for the Future
c/o Bioversity International
Stesen Kuarantin, Jabatan Pertanian
Bangunan JKR (P) 1746
P.O. Box 236, UPM Post Office
43400 Serdang, Selangor
Malaysia.
info@cropsforthefuture.org

The MEDUSA Network
- http://medusa.maich.gr/ - A network devoted to documenting the ethnobotanical resources of the Mediterranean Basin. The website includes a plant database.
medusa@maich.gr

CIAT (Centro Internacional de Agricultura Tropical – International Center for Tropical Agriculture)
- http://www.ciat.cgiar.org/Paginas/index.aspx -
Improves eco-efficient agriculture to help alleviate hunger and poverty in tropical regions.
Km 17, Recta Cali-Palmira
Apartado Aéreo 6713
Cali, Colombia

Phone: +57 2 4450000 (direct), +1 650 8336625 (via USA)
Fax: +57 2 4450073 (direct), +1 650 8336626 (via USA)
E-mail: ciat@cgiar.org

SATG (Somali Agricultural Technical Group)
- http://www.satg.org/ - An organization dedicated to helping establish a local food supply in Somalia through sustainable agricultural development.
Wendy Court Office Park (Unit 3)
David Osieli Road/Westlands Avenue
Nairobi, Kenya
P.O. Box 66286
Nairobi 00800
Dr. Hussein Haji
+254 713 970 351

MARAMAII
- http://www.marama.life.ku.dk/ - Dedicated to improving the Marama Bean (*Tylosema esculentum*).
Coordinator: Aase Hansen
University of Copenhagen
Faculty of Life Sciences
Department of Food Science
Rolighedsvej 30
DK-1958 Frederiksberg C
Denmark
E-mail: aah@life.ku.dk
Tel: +45 35 33 32 41
Fax: +45 35 33 32 45

ECHO (Educational Concerns for Hunger Organization)
- http://www.echonet.org/ - Dedicated to ending world hunger by educating and assisting aid workers. Their website contains a nice online bookstore. Provides small accessions of seeds for aid workers.
17391 Durrance Road
N. Fort Myers, FL 33917
info@echonet.org

KOMAZA
- http://www.komaza.com/ - Partners with families in Africa's semi-arid regions to produce small-scale tree farms, and productive farms.
5835 Doyle St, Ste 103
Emeryville, CA 94608
+ (415) 513-5787
engage@komaza.org

Eden Foundation
- http://www.eden-foundation.org/index.html - Works in the Sahelian region of Niger to plant multi-species, perennial forests containing useful and edible plants, in order to relieve famine and malnutrition.
Eden Foundation
Skreavägen 45B
311 72 Falkenberg
Sweden
+46-346 53157
email: project@edenfoundation.org

Recommended Reading:

Rainwater Harvesting for Drylands (Volumes 1-3) – Brad Lancaster
West Coast Food Forestry – Rain Tenaqiya
Perennial Vegetables – Eric Toensmeier
A Cactus Odyssey: Journeys in the Wilds of Bolivia, Peru, and Argentina – James D. Mauseth, Roberto Kiesling, Carlos Ostolaza
Imperfect Balance: Landscape Transformations in the Precolumbian Americas – David L. Lentz
Invasion Biology: Critique of a Pseudoscience – David I. Theodoropoulos
Ancient Agriculture: Roots and Application of Sustainable Farming – Gabriel Alonso De Herrera, Bryan Romero and Juan Estevan Arellano
Lost Crops of Africa (Volumes 1-3) – National Research Council
Lost Crops of the Incas – National Research Council
Non-Timber Uses of Selected Arid Zone Trees and Shrubs in Africa – F.E.M. Booth and G.E. Wickens
People's plants: a guide to useful plants of southern Africa – Ben-Erik van Wyk, Nigel Gericke

Resources

Tree Crops: A Permanent Agriculture – J. Russell Smith
Amaranth to Zai Holes: Ideas for Growing Food Under Difficult Conditions – Laura S. Metzner, Martin L. Price, Christi Sobel
Water for every farm using the Keyline Plan – P.A. Yeomans

Bibliography

Introduction

1. "Increasing Demand and Climate Change Threatening World Water Resources Says New UN World Water Development Report." *Unesco.org.* United Nations Educational, Scientific and Cultural Organization, 12 Mar. 2012. Accessed: 19 May 2012.

2. Power, Matthew. "Peak Water: Aquifers and Rivers Are Running Dry. How Three Regions Are Coping." *Wired Magazine* 21 Apr. 2008.

3. "Water Statistics - Water Use." *UN-Water Statistics - Water Resources.* Accessed: 19 May 2012. <http://www.unwater.org/>.

4. "Water Woes: A Global Shortage." *The Why Files.* 26 Apr. 2001. Accessed: 19 May 2012. <http://whyfiles.org/>.

5. "The Ogallala Aquifer Is Drying Up Which Means The Great Plains Could Soon Turn Into The Great American Desert." *The Final Hour.* 30 Apr. 2010. Accessed: 19 May 2012. <http://thefinalhour.blogspot.com/>.

6. Laurence, Charles. "US Farmers Fear the Return of the Dust Bowl." *Telegraph.co.uk.* The Telegraph, 7 Mar. 2011. Accessed: 19 May 2012.

7. "Water Is Key to Food Security." *FAO Media Centre.* Food and Agriculture Organization of the United Nations, 23 Aug. 2011. Accessed: 19 May 2012. <http://www.fao.org/>.

8. "Corn Ethanol Takes More Energy Than It Makes." *Environment News Service - International Daily Newswire.* 13 July 2005. Accessed: 19 May 2012. <http://www.ens-newswire.com/>.

9. Diep, Francie. "Lawns vs. Crops in the Continental U.S." *Scienceline.* New York University, 3 July 2011. Accessed: 19 May 2012. <http://scienceline.org/>.

10. Sanburn, Josh. "Scientists Create 52 Artificial Rain Storms in Abu Dhabi Desert." *TIME Newsfeed.* Time, Inc., 3 Jan. 2011. Accessed: 19 May 2012. <http://newsfeed.time.com/>.

11. Garvi, Arne. "The Lost Treasures of Eden." *Eden Foundation.* Dec. 2006. Accessed: 19 May 2012. <http://www.eden-foundation.org/>.

12. United Nations. "Desertification, Drought and Climate Change." *Trends in Sustainable Development – Africa Report* (2008-2009): 39-45.

13. Hoagland, Richard C., and David Wilcock. "Interplanetary 'Day After Tomorrow?'" *Enterprise Mission*. 14 May 2004. Accessed: 19 May 2012. <http://www.enterprisemission.com/>.

14. Pollan, Michael. "Unhappy Meals." *The New York Times Magazine*. 28 Jan. 2007.

Elegant Wattle (*Acacia victoriae*)

Bushfood Australia. www.bushfoodaustralia.com.au (accessed January, 2012).

Brand, J.C. & Maggiore, P. The nutritional composition of Australian Acacia seeds, in A.P.N.House & C.E.Harwood (eds), *Australian Dry-zone Acacias for Human Food*, pp. 54-67. Australian Tree Seed Centre, CSIRO Division of Forestry: Canberra, 1992.

Cossalter, C. Introducing Australian acacias in dry, tropical Africa, in J.W.Turnbull ed. *Australian Acacias in Developing Countries*. ACIAR Proceedings No.16, pp.118-125. Australian Centre for International Agricultural Research: Canberra, 1987.

Harwood, C.E. Human food potential of the seeds of some Australian dry-zone Acacia species, *Journal of Arid Environments* 27: 27-35, 1994.

Harwood, C.E., Rinaudo, T. & Adewusi, S. Developing Australian acacia seeds as a human food for the Sahel, *Unasylva* 196: 57-64, 1999.

Latz, P.K. *Bushfires and Bushtuckers: Aboriginal Plant Use in Central Australia*. IAD Press: Alice Springs, 1995.

Maslin, B.R., Thomson, L.A.J., McDonald, M.W. and Hamilton-Brown, S. *Edible wattle seeds of southern Australia: A review of species for use in semi-arid regions*. CSIRO: Australia, 1998.

Boland, D.J., Brooker, M.I.H., Chippendale, G.M., Hall, N., Hyland, B.P.M., Johnson, R.D., Kleinig, D.A., McDonald, M.W., Turner, J.D. *Forest Trees of Australia*. CSIRO Publishing: Collingwood, 2006.

Doran JC, Turnbull J.W. (eds.) *Australian Trees and Shrubs: species for land rehabilitation and farm planting in the tropics*. Australian Centre for International Agricultural Research: Canberra, 1997.

Cunningham, G.M., Mulham, W.E., Milthorpe, P.L., Leigh, J.H. *Plants of Western New South Wales*, Inkata Press: Port Melbourne, Victoria, 1992.

Rooibos (*Aspalathus linearis*)

Natural Standard Corporation. "Rooibos (*Aspalathus linearis*) Plant Profiler." *Sigma-Aldrich*. Natural Standard, 2010. 29 Jan. 2012. <http://www.sigmaaldrich.com/>.

Gerard, Anne. Habitat conditions of wild Rooibos tea (*Aspalathus linearis*): Environmental abiotic and biotic drivers of its performance - BSc Thesis, University of Hamburg, 2010.

Govender, Marcini. "Aspalathus linearis." *PlantZAfrica.com* Homepage. National Herbarium, June 2007. (accessed 3 April 2011). <http://www.plantzafrica.com/>.

Pigeon Pea (*Cajanus cajan*)

James A. Duke. 1983. *Handbook of Energy Crops*. unpublished.

Duke, J.A. 1981. *Handbook of legumes of world economic importance*. Plenum Press. NewYork.

Duke, J.A. 1981b. The gene revolution. Paper 1. p. 89–150. In: Office of Technology Assessment, Background papers for innovative biological technologies for lesser developed countries. USGPO. Washington.

Morton, J.F. 1976. The pigeon pea (Cajanus cajan Millsp.), a high protein tropical bush legume. HortScience 11(1):11–19.

N.A.S. 1980a. *Firewood crops: Shrub and tree species for energy production*. National Academy of Sciences, Washington, DC.

Le Houérou, Henri. "Cajanus Cajan." *FAO: Food and Agriculture Organization of the United Nations*. Accessed: 01 Feb. 2012. <http://www.fao.org/>.

Caper (*Capparis spinosa*)

Fici, S (October 2001). "Intraspecific variation and evolutionary trends in *Capparis spinosa* L. (Capparaceae)". *Plant Systematics and Evolution* (Springer Wien) 228 (3-4): 123–141.

Inocencio, C; F Alcaraz, F Calderón, C Obón, D Rivera (April, 2002). "*The use of floral characters in Capparis sect. Capparis' to determine the botanical and geographical origin of capers*". European Food Research and Technology (*Springer*) 214 (4): 335–339.

Andrade, G; E Esteban, L Velasco, MJ Lorite, EJ Bedmar (1997). "Isolation and identification of N2-fixing microorganisms from the rhizosphere of *Capparis spinosa* (L.).". *Plant and Soil* (Kluwer Academic Publishers) 197 (1): 19–23.

Sozzi, GO; A Chiesa (June 1995). "Improvement of caper (*Capparis spinosa* L.) seed germination by breaking seed coat-induced dormancy.". *Scientia Horticulturae* (Elsevier) 62 (4): 255–261.

Luna Lorente, F; M Pérez Vicente (1985). La Tapenera o Alcaparra: Cultivo y Aprovechamiento. Publicaciones de Extensión Agraria, Colección Agricultura Práctica 37. Ministerio de Agricultura, Pesca y Alimentación, Madrid, España, 125 pp.

Alkire, Ben. "Caper - Capparis Spinosa." *Center for New Crops and Plant Products*. Purdue University, 1998. Accessed: 05 Feb. 2012. <http://www.hort.purdue.edu/newcrop/cropfactsheets/caper.html>.

Carob (*Ceratonia siliqua*)

El Idrissi, M. Missbah, N. Aujjar, A. Belabed, Y. Dessaux, and A. Filali-Maltouf. "Characterization of Rhizobia Isolated from Carob Tree (*Ceratonia siliqua*)." *Journal of Applied Microbiology* 80.2 (1996): 165-73.

Morton, Julia Frances, and Curtis F. Dowling. *Fruits of Warm Climates*. Miami, FL: J.F. Morton, 1987.

Anon. 1986. *The useful plants of India*. Publications & Information Directorate, CSIR, New Delhi, India.

Batlle I, Tous J. Carob tree, Ceratonia siliqua L. Promoting the conservation and use of underutilized and neglected crops. No. 17. Institute of Plant Genetics and Crop Plant Research, Gatersleben/International Plant Genetic Resources Institute, Rome.

Esbenshade HW, Wilson G. 1986. *Growing carobs in Australia*. Goddard and Dobson Publishers. Victoria.

Hong TD, Linington S, Ellis RH. 1996. Seed storage behaviour: a compendium. Handbooks for Genebanks: No. 4. IPGRI.

Lanzara P and Pizetti M. 1978. *Simon & Schuster's guide to trees*. Simon & Schuster, Inc., New York.

Tropical Legumes: Resources for the Future. Washington, D.C.: National Academy of Science, 1977.

Yeheb Nut (*Cordeauxia edulis*)

Booth, F. E. M., and G. E. Wickens. *Non-timber Uses of Selected Arid Zone Trees and Shrubs in Africa*. Rome: Food and Agriculture Organization of the United Nations, 1988.

Tropical Legumes: Resources for the Future. Washington, D.C.: National Academy of Science, 1977.

Söderberg, Vera. "The Importance of Yeheb (*Cordeauxia edulis*) for Somali Livestock Production and Its Effects on Body Tissues When Fed to Swedish Domestic Goats." *Department of Anatomy, Physiology and Biochemistry, Swedish University of Agricultural Sciences*.

Honeybush (*Cyclopia genistoides*)

Van Der Walt, Liesl. "Cyclopia genistoides." *PlantZAfrica.com Homepage*. Sept. 2000. Accessed: 18 Feb. 2012. <http://www.plantzafrica.com/>.

"Honeybush Tea - Organic Honeybush Tea Is Caffeine Free." *Rooibos Tea, Organic Rooibos Tea & Green Rooibos Tea, Bulk, Wholesale and Retail*. Accessed: 18 Feb. 2012. <http://www.montegotea.com/>.

Goumi (*Elaeagnus multiflora*)

Lee, Mee Sook, Yun-Kyoung Lee, and Ock Jin Park. "Cherry Silver Berry (Elaeagnus multiflora) Extracts Exert Antiinflammatory Effects by Inhibiting COX-2 and Akt Signals in HT-29 Colon Cancer Cells." *Food Science and Biotechnology* 19.6 (2010): 1673-677.

Baker, Dwight. "Elaeagnus: A Widely Distributed Temperate Nitrogen Fixer." *Elaeagnus*. Forest, Farm, and Community Tree Network (FACT Net), 1992. Accessed: 09 Mar. 2012. <http://www.winrock.org/>.

Corleto, A., E. Cazzato, and V. Laudadio. "Quantitative and Qualitative Evaluation of Tree and Shrubby Pasture Species in Southern Italy." *Options Méditerranéennes* 4 (1994): 129-34.

Plants for a Future Database. www.pfaf.org (accessed March, 2012).

African Winterthorn (*Faidherbia albida*)

Booth, F. E. M., and G. E. Wickens. *Non-timber Uses of Selected Arid Zone Trees and Shrubs in Africa*. Rome: Food and Agriculture Organization of the United Nations, 1988.

"*Faidherbia albida*: Direct Seeding Experiments in an Arid Environment." *Eden Foundation*. Apr. 1991. Accessed: 25 Feb. 2012.
<http://www.eden- foundation.org/>.

Fagg, C. W. "*Faidherbia albida* - Inverted Phenology Supports Dryzone Agroforestry." *NFT Highlights*. Forest, Farm, and Community Tree Network (FACT Net), Jan. 1995. Accessed: 25 Feb. 2012. <http://www.winrock.org/>.

Honey Locust (*Gleditsia triacanthos*)

Bryan, James A., Graeme P. Berllyn and John C. Gordon. 1996. Toward a new concept of the evolution of symbiotic nitrogen fixation in the Leguminosae. Plant and Soil. 186:151-159.

Duke, James A., and Edward S. Ayensu. *Medicinal Plants of China*. Algonac, MI: Reference Publications, 1985.

Blair, Robert M. "*Gleditsia triacanthos* L." *Northeastern Area State & Private Forestry*. Accessed: 07 Mar. 2012. <http://www.na.fs.fed.us/>.

Gold, Michael A. "*Gleditsia triacanthos* - Honeylocust, Widely Adapted Temperate Zone Fodder Tree." *Winrock International*. Forest, Farm, and Community Tree Network (FACT Net). Accessed: 07 Mar. 2012. <http://www.winrock.org/>.

False Mopane (*Guibourtia coleosperma*)

The Wood Explorer Database. Accessed: 11 Mar. 2012. <http://www.thewoodexplorer.com/>.

"*Guibourtia coleosperma*." *Flora of Zimbabwe: Species Information*. Accessed: 17 Apr. 2011. <http://www.zimbabweflora.co.zw/>.

"*Guibourtia coleosperma* (Copalwood)." *Biodiversity Explorer*. Accessed: 17 Apr. 2011. <http://www.biodiversityexplorer.org/>.

Leucaena (*Leucaena leucocephala*)

Gutteridge, Ross C., and H. Max. Shelton. *Forage Tree Legumes in Tropical Agriculture*. Wallingford, Oxon: CAB International, 1994.

Duke, J.A. *Handbook of Energy Crops*. 1983

"*Leucaena leucocephala*." *Tropical Forages*. Accessed: 19 Mar. 2012. <http://www.tropicalforages.info/>.

Brewbaker, J. L. "*Leucaena leucocephala* - a Versatile Nitrogen Fixing Tree." *Winrock International*. Forest, Farm, and Community Tree Network (FACT Net). Accessed: 19 Mar. 2012. <http://www.winrock.org/>.

Blue Lupin (*Lupinus angustifolius*)

"Australian Sweet Lupin (*Lupinus angustifolius*)." *Department of Agriculture and Food: Government of Western Australia.* Accessed: 19 Mar. 2012. <http://www.agric.wa.gov.au/>.

Duke, James A. *Handbook of Legumes of World Economic Importance.* New York: Plenum, 1981.

Ad Hoc Panel of the Advisory Committee on Technology Innovation, Board on Science and Technology for International Development, National Research Council. *Lost Crops of the Incas: Little-Known Plants of the Andes with Promise for Worldwide Cultivation.* Washington D.C.: National Academies, 1989.

Martins, J. M. "Cholesterol-lowering Effects of Dietary Blue Lupin (*Lupinus angustifolius* L.) in Intact and Ileorectal Anastomosed Pigs." *The Journal of Lipid Research* 46.7 (2005): 1539-547.

Plants for a Future Database. www.pfaf.org (accessed January, 2012).

Tepary Bean (*Phaseolus acutifolius*)

Hamama, A.A. and H.L. Bhardwaj. 2002. Tepary bean: A short duration summer crop in Virginia. p. 429–431. In: J. Janick and A. Whipkey (eds.), Trends in new crops and new uses. ASHS Press, Alexandria, VA.

Tropical Legumes: Resources for the Future. Washington, D.C.: National Academy of Science, 1977.

Stephens, James M. "Bean, Tepary — *Phaseolus acutifolius* A. Gray." *University of Florida IFAS Extension.* Horticultural Sciences Department, Florida Cooperative Extension Service, Apr. 1994. Accessed: 7 Apr. 2012. <http://edis.ifas.ufl.edu/mv025>.

"*Phaseolus acutifolius* (Pinacate Tepary Bean)." *Backyard Gardener.* Accessed: 07 Apr. 2012. <http://www.backyardgardener.com/>.

Manila Tamarind (*Pithecellobium dulce*)

Subhadrabandhu, Suranant. *Under-utilized Tropical Fruits of Thailand.* Bangkok: FAO Regional Office for Asia and the Pacific, 2001.

Allen, O.N., and E.K. Allen. *The leguminosae: a source book of characteristics, uses and modulation.* Wisconsin: Wisconsin Press, 1981.

Ambasta, Shri S.P, ed. *The useful plants of India.* New Delhi, India: Council of Scientific and Industrial Research, 1986.

National Academy of Sciences. *Firewood crops: shrub and tree species for energy production.* Washington, DC: NAS/NRC, 1980.

Little, E.L. *Common fuelwood crops.* Morgantown, WV: Communi-Tech Association, 1985.

Little, E.L. Jr., and F.H. Wadsworth. "Common trees of Puerto Rico and the Virgin Islands." Washington, DC: *USDA Forest Service Agricultural Handbook* no. 249, 1964.

Argentine Mesquite (*Prosopis alba*)

Little EL. 1983. *Common fuelwood crops.* Communi-Tech Association, Morgantown, West Virginia.

MacDicken GK. 1994. Selection and management of nitrogen fixing trees. Winrock International, and Bangkok: FAO.

National Academy of Sciences. 1980. *Firewood crops.* National Academy Press. Washington D.C.

NFTA. 1989. NFT gums: ancient and modern commercial products. NFTA 89-01. Waimanalo.

NFTA. 1991. Prosopis alba and Prosopis chilensis: subtropical semi-arid fuel and fodder trees. NFTA 91-02. Waimanalo.

Uchytil, Ronald J. 1990. *Prosopis velutina.* In: Fire Effects Information System, [Online]. U.S. Department of Agriculture, Forest Service, Rocky Mountain Research Station, Fire Sciences Laboratory Producer. Accessed: 15 Apr 2012. <http://www.fs.fed.us/database/feis/>

Ewens M., Gezan S., Felker P. Five Year Field Evaluation of *Prosopis alba* Clones on pH 9–10 Soils in Argentina Selected for Growth in the Greenhouse at Seawater Salinities (45 dS m−1). *Forests.* 2012; 3(1):95-113.

Felker, P.; Lopez, C.; Soulier, C.; Ochoa, J.; Abdala, R.; Ewens, M. Genetic evaluation of *Prosopis alba* (algarrobo) in Argentina for cloning elite trees. *Agroforestry Systems.* 2001; 53 (1): 65-76.

Duke, James A. 1983. *Handbook of Energy Crops.* 1983.

World Wildlife Fund (Lead Author);Mark McGinley (Topic Editor) "Arid Chaco". In: Encyclopedia of Earth. Eds. Cutler J. Cleveland (Washington, D.C.: Environmental Information Coalition, National Council for Science and the Environment). [First published in the Encyclopedia of Earth March 6, 2007; Last revised Date March 6, 2007; Retrieved April 16, 2012. <http://www.eoearth.org/>.

Karoo Boer Bean (*Schotia afra var. afra*)

Oliver, Ian. "Schotia afra var. afra." *PlantZAfrica.com Homepage*. Karoo Desert National Botanic Garden, Mar. 2003. Accessed: 26 Apr. 2012. <http://www.plantzafrica.com/>.

Mbambezeli, Giles, and Alice Notten. "*Schotia brachypetala*." *PlantZAfrica.com Homepage*. Kirstenbosch National Botanic Garden, Aug. 2001. Accessed: 26 Apr. 2012. <http://www.plantzafrica.com/>.

Tamarind (*Tamarindus indica*)

Morton, Julia F. *Fruits of Warm Climates*. Miami, FL, 1987.

Janick, Jules, and Robert E. Paull. *The Encyclopedia of Fruit and Nuts*. Cambridge, MA: CABI North American Office, 2006.

Hines, Deborah A., and Karlyn Eckman. *Indigenous Multipurpose Trees of Tanzania: Uses and Economic Benefits for People*. Canada: Cultural Survival Canada, 1993.

Tropical Legumes: Resources for the Future. Washington, D.C.: National Academy of Science, 1977.

Marama Bean (*Tylosema esculentum*)

van der Maesen, L.J.G., 2006. *Tylosema esculentum* (Burch.) A.Schreib. Record from Protabase. Brink, M. & Belay, G. (Editors). PROTA (Plant Resources of Tropical Africa / Ressources végétales de l'Afrique tropicale), Wageningen, Netherlands. <http://database.prota.org/>. Accessed 10 May 2012.

Powell, Michael A. "Marama Bean (*Tylosema esculentum*, Fabaceae) Seed Crop in Texas." *Economic Botany* 41.2 (1987): 216-20.

Lost Crops of Africa: Volume II: Vegetables. Washington, D.C.: National Academies, 2006.

Moth Bean (*Vigna aconitifolia*)

Brink, M. & Jansen, P.C.M., 2006. *Vigna aconitifolia* (Jacq.) Maréchal. Record from Protabase. Brink, M. & Belay, G. (Editors). *PROTA (Plant Resources of Tropical Africa / Ressources végétales de l'Afrique tropicale)*, Wageningen, Netherlands. <http://database.prota.org/>. Accessed 11 May 2012.

Tropical Legumes: Resources for the Future. Washington, D.C.: National Academy of Science, 1977.

Narain, P., Singh, R.S. & Kumar, D. “Droughts and dew bean productivity in northwestern arid Rajasthan, India.” *Drought Network News* 13.1 (2000): 7–9.

Sathe, S., and M. Venkatachalam. "Fractionation and Biochemical Characterization of Moth Bean (*Vigna aconitifolia* L.) Proteins." *LWT - Food Science and Technology* 40.4 (2007): 600-610.

Bambara Groundnut (*Voandzeia subterranea*)

Brink, M., Ramolemana, G.M. & Sibuga, K.P., 2006. *Vigna subterranea* (L.) Verdc. Record from Protabase. Brink, M. & Belay, G. (Editors). PROTA (Plant Resources of Tropical Africa / Ressources végétales de l'Afrique tropicale), Wageningen, Netherlands. <http://database.prota.org/>. Accessed 12 May 2012.

Lost Crops of Africa: Volume II: Vegetables. Washington, D.C.: National Academies, 2006.

"*Vigna subterranea*." Ecocrop. FAO - Food and Agriculture Organization of the UN. Accessed 12 May 2012. <http://ecocrop.fao.org/>.

Bamshaiye, O.M., J.A Adegbola, and E.I. Bamishaiye. "Bambara Groundnut: An Under-Utilized Nut in Africa." *Advances in Agricultural Biotechnology* 1 (2011): 60-72.

Jelly Bean Palm (*Butia capitata*)

"Pindo Palms - (*Butia capitata*) - Cold Hardy Palm Trees." *Palm Tree Types and Palm Tree Pictures from Sun Palm Trees.* Accessed: 03 April 2011. <http://www.sunpalmtrees.com/>.

"Butia capitata." *UA Campus Arboretum.* Accessed: 03 Feb. 2012. <http://arboretum.arizona.edu/>.

Pereira Faria, Juliana, Egle M.A. Siquiera, Roberto Fontes Vieira, and Tania Da Silveira Agostini-Costa. "Fruits of *Butia capitata* (Mart.) Becc as Good Sources of Beta-Carotene and Provitamin A." *Rev. Bras. Frutic., Jaboticabal - SP* Volume Especial, E (2011): 612-17.

Gingerbread Palm (*Hyphaene thebaica*)

Soromessa, T., 2011. Hyphaene thebaica (L.) Mart. Record from PROTA4U. Brink, M. & Achigan-Dako, E.G. (Editors). PROTA (Plant Resources of Tropical Africa / Ressources végétales de l'Afrique tropicale), Wageningen, Netherlands. <http://www.prota4u.org/>. Accessed 12 March 2012 .

"*Hyphaene thebaica.*" *World Agroforestry Centre.* 13 Mar. 2012. <http://www.worldagroforestry.org/>.

"View Crop: *Hyphaene thebaica.*" *Ecocrop.* Food and Agriculture Organization of the UN. 13 Mar. 2012. <http://ecocrop.fao.org/>.

Blach-Overgaard, A., J-C Svenning, and H. Balslev. "Climate Change Sensitivity of the African Ivory Nut Palm,Klotzsch Ex Mart. (Arecaceae) – a Keystone Species in SE Africa." *IOP Conference Series: Earth and Environmental Science* 8 (2009).

Chilean Wine Palm (*Jubaea chilensis*)

"Coquito Nuts." *Frieda's The Specialty Produce People.* Accessed: 13 Mar. 2012. <http://www.friedas.com/>.

Sturtevant, E. Lewis. *Sturtevant's Edible Plants of the World.* Ed. U. P. Hedrick. New York: Dover Publications, 1972.

Hoffmann-J., Adriana E., and Olga Herrera-MacBryde. "Mediterranean Region And La Campana National Park Central Chile." *Botany Department, Smithsonian Institution.* 13 Mar. 2012. <http://botany.si.edu/>.

Campos, Roasio Jorge. *Productos Forestales No Madereros En Chile.* Santiago, Chile: Oficina Regional De La FAO Para América Latina Y El Caribe, 1998.

"*Jubaea chilensis.*" *Chilebosque.* 13 Mar. 2012. <http://www.chilebosque.cl/tree/jchil.html>.

Bergman, Phil. "*Jubaea chilensis* The Chilean Wine Palm." Jungle Music Palms and Cycads. Accessed: 13 Mar. 2012. <http://www.junglemusic.net/>.

Mountain Coconut (*Parajubaea torallyi var. torallyi*)

Lost Crops of the Incas: Little-known Plants of the Andes with Promise for Worldwide Cultivation. Washington, D.C.: National Academy, 1989.

"PACSOA - *Parajubaea torallyi.*" *Palm and Cycad Societies of Australia.* Accessed: 03 Apr. 2011. <http://www.pacsoa.org.au/>.

Haynes, Jody, and John McLaughlin. "Edible Palms and Their Uses." *Quisqualis Rare Fruit and Plant Information.* Tropical Visions, May 2004. Accessed: 02 Apr. 2012. <http://www.quisqualis.com/>.

"*Parajubaea torallyi.*" *Botanix.* KPR, 7 May 2009. 02 Apr. 2012. <http://www.botanix.kpr.eu/>.

"Guide to Species *Parajubaea torallyi torallyi*." *Palmeras Y Jardines*. Palmasur. 2 Apr. 2012. <http://www.palmerasyjardines.com/>.

"*Parajubaea torallyi var. torallyi*." *AMAZÒNIA Files*. Accessed: 02 Apr. 2012. <http://www.amazonia-online.com/>.

Date Palm (*Phoenix dactylifera*)

Morton, Julia F. *Fruits of Warm Climates*. Miami, FL, 1987.

Zaid, A., and E. J. Arias-Jiménez. *Date Palm Cultivation*. Rome: Food and Agriculture Organization of the United Nations, 2002.

"*Phoenix Dactylifera*." *World Agroforestry Centre*. Accessed: 09 Apr. 2012. <http://www.worldagroforestrycentre.org/>.

Armstrong, W. P. "Date Palm (*Phoenix Dactylifera*)." *WAYNE'S WORD*. Accessed: 09 Apr. 2012. <http://waynesword.palomar.edu/>.

Haynes, Jody, and John McLaughlin. "Edible Palms and Their Uses." *Quisqualis Rare Fruit and Plant Information*. Tropical Visions, May 2004. Accessed: 09 Apr. 2012. <http://www.quisqualis.com/>.

Licuri Palm (*Syagrus coronata*)

Crepaldi, Iara Cândido, Ligia Bicudo De Almeida-Muradian, Magda Dias Gonçalves Rios, Marilene De Vuono Camargo Penteado, and Antonio Salatino. "Composição Nutricional Do Fruto De Licuri (*Syagrus coronata* (Martius) Beccari)." *Revista Brasileira De Botânica* 24.2 (2001): 155-159.

Crepaldi, Iara Cândido, Antonio Salatino, and Alone Rios. "*Syagrus coronata* and *Syagrus vagans*: Traditional Exploitation in Bahia, Brazil." *Palms* 48.1 (2004): 43-48.

Plotkin, Mark J., and Michael J. Balick. "Medicinal Uses of South American Palms." *Journal of Ethnopharmacology* 10.2 (1984): 157-79.

ECO-CHARCOAL Project Concept Study. Hauser Estratégia & Participação Ltda., July 2010. (Accessed: 6 May 2012). <http://www.hepbrasil.com/>.

Aloe Vera (*Aloe vera*)

Shukla, Shubhlakshmi. "Aloe Vera Has Biodiesel Potential, Reveals MSU Study." *Express India*. The Indian Express Limited, 2008.

Oudhia, Pankaj. "Ghrita kumari or Guar patha." Raipur, India: *Society for Parthenium Management*, 2001.

Bibliography

Plants for a Future Database. www.pfaf.org (accessed January, 2012).

Center for New Crops & Plant Products. www.hort.purdue.edu/newcrop (accessed January, 2012).

King, G.K., Yates, K.M., Greenlee P.G., et al. "The effect of Acemannan Immunostimulant in combination with surgery and radiation therapy on spontaneous canine and feline fibrosarcomas". *Journal of the American Animal Hospital Association* 31: 439–47, 1995.

Lyons, G. "The Definitive *Aloe vera*, vera?". Huntington Botanic Gardens. (accessed April, 2011).

Darokar, M.P., Rai, R, Gupta, A.K., Shasany, A.K., Rajkumar, S., Sunderasan, V and Khanuja S.P.S. "Molecular assessment of germplasm diversity in *Aloe spp.* using RAPD and AFLP analysis." *Journal of Medicinal and Aromatic Plant Sciences* 25: 354–361, 2003.

Treutlein, J., Smith, G. F. S., van Wyk, B. E. & Wink, W. Phylogenetic relationships in Asphodelaceae (Alooideae) inferred from chloroplast DNA sequences (rbcl, matK) and from genomic finger-printing (ISSR). *Taxon* 52:193, 2003.

Jones, Warren D., and Charles M. Sacamano. *Landscape Plants for Dry Regions: More than 600 Species from around the World.* Tucson, AZ: Fisher, 2000.

Cosmetic Ingredient Review Expert Panel. "Final report on the safety assessment of AloeAndongensis Extract, Aloe Andongensis Leaf Juice,aloe Arborescens Leaf Extract, Aloe Arborescens Leaf Juice, Aloe Arborescens Leaf Protoplasts, Aloe Barbadensis Flower Extract, Aloe Barbadensis Leaf, Aloe Barbadensis Leaf Extract, Aloe Barbadensis Leaf Juice,aloe Barbadensis Leaf Polysaccharides, Aloe Barbadensis Leaf Water, Aloe Ferox Leaf Extract, Aloe Ferox Leaf Juice, and Aloe Ferox Leaf Juice Extract". *International Journal of Toxicology* 26: 1–50, 2007.

Maenthaisong, R., Chaiyakunapruk, N., Niruntraporn, S. et al. "The efficacy of aloe vera for burn wound healing: a systematic review". *Burns* 33 (6): 713–718, 2007.

Feily, A., Namazi, M.R. "*Aloe vera* in dermatology: a brief review". *G Ital Dermatol Venereol* 144: 84–91, 2009.

Sumbul, S., Waseemuddin, A.S., Iqbal, A. "Antifungal activity of *Allium*, *Aloe*, and *Solanum* species". *Pharmaceutical Biology* 42 (7): 491–498, 2004.

Apple Cactus (*Cereus repandus*)

J. Weiss, A. Nerd, and Y. Mizrahi. 1993. Development of *Cereus peruvianus* (Apple Cactus) as a new crop for the Negev desert of Israel. p. 486-471. In: J. Janick and J.E. Simon (eds.), New crops. Wiley, New York.

Kalahari Cucumber (*Coccinia sessilifolia*)

Bosch, C.H., 2004. "*Coccinia sessilifolia*" (Sond.) Cogn. In: Grubben, G.J.H. & Denton, O.A. (Editors). PROTA 2: Vegetables/Légumes. PROTA, Wageningen, Netherlands.

Sacred Succulents. Accessed: 16 Feb. 2012. <http://sacredsucculents.com/>.

Sanky (*Corryocactus brevistylus*)

Matos-Chamorro, Alfredo, Julio Paredes-Guzmán, and Luisa González-Rengifo. "Determinación De La Capacidad Antioxidante De Los Compuestos Fenólicos Del Sancayo (*Corryocactus brevistylus*)." *Revista De Investigación En Ciencia Y Tecnología De Alimentos* 1.1 (2010): 66-71. Dirección General De Investigación.

Peru Natural Sanky. Accessed: 22 Feb. 2012. <http://www.sanky.galeon.com/>.

"Description and Images of *Corryocactus brevistylus* (Guacalla), a Native Chilean Plant, Provided by the Supplier of Native Exotic Chilean Seeds, Chileflora.com." *Homepage of ChileFlora.com, a Supplier of Native Chilean Seeds and Seedlings and Organizer of Plant-watching Tours*. 2009. Accessed: 22 Feb. 2012. <http://www.chileflora.com/>.

Sacred Succulents. Accessed: 22 Feb. 2012. <http://sacredsucculents.com/>.

Rounded Noon-Flower (*Disphyma crassifolium subsp. clavellatum*)

Kapitany, Attila. *Australian Succulent Plants*. Boronis (Australia): Kapitany Concepts, 2007.

"*Disphyma crassifolium subsp. clavellatum* (Haw.) Chinnock: FloraBase: Flora of Western Australia." *FloraBase The Western Australian Flora*. Accessed: 3 Apr. 2011. <http://florabase.dec.wa.gov.au/>.

"Rounded Noon-flower." *Department of Primary Industries and Department of Sustainability and Environment*. The State of Victoria, 25 Mar. 2011. Accessed: 3 Apr. 2011. <http://vro.dpi.vic.gov.au/>.

Ruby Salt Bush (*Enchylaena tomentosa*)

Kapitany, Attila. *Australian Succulent Plants*. Boronis (Australia): Kapitany Concepts, 2007.

"*Enchylaena tomentosa*." *Australian Native Plants Society*. Accessed: 3 Apr. 2011. <http://anpsa.org.au/>.

Hadlow, Barrie. "Enchylaena Tomentosa." - *Growing Native Plants*. Australian National Botanic Gardens, 1986. Accessed: 26 Feb. 2012. <http://www.anbg.gov.au/>.

Midnight Lady (*Harrisia pomanensis*)

Gordillo, Sandra. *Plantas Nativas De Cerro Colorado*. 2009. Accessed: 12 Mar. 2012. <http://www.cerrocoloradoweb.com.ar/>.

World Wildlife Fund (Lead Author); Mark McGinley (Topic Editor) "Arid Chaco". In: Encyclopedia of Earth. Eds. Cutler J. Cleveland (Washington, D.C.: Environmental Information Coalition, National Council for Science and the Environment). First published in the Encyclopedia of Earth March 6, 2007; Last revised Date March 6, 2007; Retrieved April 16, 2012. <http://www.eoearth.org/>.

"Harrisia Cactus." *Weeds of Australia: Biosecurity Queensland Edition*. The University of Queensland, 2011. Accessed: 12 Mar. 2012. <http://keyserver.lucidcentral.org/>.

Ice Plant (*Mesembryanthemum crystallinum*)

Sturtevant, E. Lewis. *Sturtevant's Edible Plants of the World*. Ed. U. P. Hedrick. New York: Dover Publications, 1972.

Plants for a Future Database. www.pfaf.org (accessed April, 2011).

Fara, Heather. "Ice Plant, Iceplant." *The Floral Genome Project*. University of Florida. Accessed: 22 Mar. 2012. <http://www.flmnh.ufl.edu/>.

"*Mesembryanthemum Spp.*" *Australian National Botanic Gardens*. Accessed: 22 Mar. 2012. <http://www.anbg.gov.au/cpbr/WfHC/Mesembryanthemum/>.

Smith, Michael T., Neil R. Crouch, Nigel Gericke, and Manton Hirst. "Psychoactive Constituents of the Genus Sceletium N.E.Br. and Other Mesembryanthemaceae: A Review." *Journal of Ethnopharmacology* 50.3 (1996): 119-30.

Indian Fig (*Opuntia ficus-indica*)

Bein E. 1996. *Useful trees and shrubs in Eritrea*. Regional Soil Conservation Unit (RSCU), Nairobi, Kenya.

Brutsch MO and HelmuthGZ. 1990. *The pricky pear (Opuntia ficus-indica[Cactaceae]) in South Africa: utilization of the naturalized weed and of the cultivated plants*. Economic Botany 47(2): 154-162.

FAO. 1988. *Traditional food plants: a resource book for promoting the exploitation and consumption of food plants in arid, semi-arid and sub-humid lands of East Africa*. FAO food and nutrition paper 42. FAO, Rome.

Le Houerou HN. 1994. *The role of cacti (Opuntia spp.) in erosion control, land reclamation, rehabilitation and agricultural development in the Mediterranean Basin*. Journal of Arid Environments 33: 135-159.

Russel CE. 1988. *Stabilizing productivity of semi-arid regions: the case for cactus/woody legume silvopastoral systems*. Paper presented at the international range land Development symposium. Corpus Christi, Texas, Feb 25, 1988.

Plants for a Future Database. www.pfaf.org (accessed April, 2011).

Salicornia (*Salicornia bigelovii*)

Lu, Donghe, Min Zhang, Shaojin Wang, Jinlong Cai, Xiang Zhou, and Chengpei Zhu. "Nutritional Characterization and Changes in Quality of *Salicornia bigelovii* Torr. during Storage." *LWT - Food Science and Technology* 43 (2009): 519-24.

Attia, F., A. Alsobayel, M. Kriadees, M. Alsaiady, and M. Bayoumi. "Nutrient Composition and Feeding Value of *Salicornia bigelovii* Torr Meal in Broiler Diets." *Animal Feed Science and Technology* 65.1-4 (1997): 257-63.

Glenn, E. P., J. W. O'Leary, M. C. Watson, T. L. Thompson, and R. O. Kuehl. "*Salicornia bigelovii* Torr.: An Oilseed Halophyte for Seawater Irrigation." *Science* 251.4997 (1991): 1065-067.

Glenn, E. "*Salicornia bigelovii* Torr.: A Seawater-irrigated Forage for Goats." *Animal Feed Science and Technology* 40.1 (1992): 21-30.

Grattan, S. R., S. E. Benes, D. W. Peters, and F. Diaz. "Feasibility of Irrigating Pickleweed (*Salicornia bigelovii*. Torr) with Hyper-saline Drainage Water." *Journal of Environmental Quality* 37.5 Supplement (2008): S-149-56.

Baldwin, Bruce G., Douglas H. Goldman, and Linda Ann. Vorobik. *The Jepson Manual: Vascular Plants of California*. Berkeley, CA: University of California, 2012.

Christiansen, Ryan C. "Sea Asparagus Can Be Oil Feedstock." *Biodiesel Magazine*. 31 July 2008. <http://www.biodieselmagazine.com/>.

"*Salicornia bigelovii* Information from NPGS/GRIN." *GRIN National Genetic Resources Program*. Accessed: 22 Apr. 2012. <http://www.ars-grin.gov/>.

"Greening of Eritrea - 17min." *YouTube*. (accessed April, 2012). <http://www.youtube.com/>.

Plants for a Future Database. www.pfaf.org (accessed April, 2012).

Kanna (*Sceletium tortuosum*)

Smith, Michael T., Neil R. Crouch, Nigel Gericke, and Manton Hirst. "Psychoactive Constituents of the Genus *Sceletium* N.E.Br. and Other Mesembryanthemaceae: A Review." *Journal of Ethnopharmacology* 50.3 (1996): 119-30.

Shikanga E. A., Hamman J.H., Chen W., Combrinck S., Gericke N., Viljoen A.M. "In vitro permeation of mesembrine-alkaloids from *Sceletium tortuosum* across porcine buccal, sublingual and intestinal mucosa". *Planta Medica*, 78 (2012): 260-268.

Patnala, Srinivas, and Isadore Kanfer. "Investigations of the Phytochemical Content of *Sceletium tortuosum* following the Preparation of "Kougoed" by Fermentation of Plant Material." *Journal of Ethnopharmacology* 121.1 (2009): 86-91.

Stafford, Gary I., Mikael E. Pedersen, Johannes Van Staden, and Anna K. Jäger. "Review on Plants with CNS-effects Used in Traditional South African Medicine against Mental Diseases." *Journal of Ethnopharmacology* 119.3 (2008): 513-37.

Chesselet, Pascale. "*Sceletium tortuosum*." *PlantZAfrica.com Homepage*. May 2005. Accessed: 25 Apr. 2012. <http://www.plantzafrica.com/>.

South African National Biodiversity Institute, South African Medical Research Council, and University of the Western Cape. "*Sceletium tortuosum* Herba." *PlantZAfrica.com*. Accessed: 25 Apr. 2012. <http://www.plantzafrica.com/>.

"*Sceletium tortuosum* - Kanna." *Entheology.org*. Accessed: 26 Apr. 2012. <http://www.entheology.org/>.

Sacred Succulents. Accessed: 25 April 2012. <http://sacredsucculents.com/>.

Galloping Cactus (*Stenocereus gummosus*)

Janick, Jules, and Robert E. Paull. *The Encyclopedia of Fruit and Nuts*. Cambridge, MA: CABI North American Office, 2006.

León de La Luz, J. L. & Domínguez, R. 1991. Evaluación de la Reproducción por semilla de la pitaya agria (*Stenocereus gummosus*) en Baja California Sur, México. *Acta Botánica Mexicana* 14:75–87.

Clark-Tapia, R., and F. Molina-Freaner. "The Genetic Structure of a Columnar Cactus with a Disjunct Distribution: *Stenocereus gummosus* in the Sonoran Desert." *Heredity* 90 (2003): 443-50.

Nerd, A., J.A. Aronson, and Y. Mizrahi. 1990. Introduction and domestication of rare and wild fruit and nut trees for desert areas. p. 355-363. In: J. Janick and J.E. Simon (eds.), Advances in new crops. Timber Press, Portland, OR.

Lauri, Bob. "SDNHM - *Stenocereus gummosus* (Sour Pitaya, Pitaya Agria, Pitahaya)." *Ocean Oasis*. Accessed: 01 May 2012. <http://www.oceanoasis.org/>.

"*Stenocereus gummosus*." *Arizona-Sonora Desert Museum Digital Library*. Accessed: 01 May 2012. <http://www.desertmuseumdigitallibrary.org/>.

"*Stenocereus queretaroensis*." *Los Diccionarios Y Las Enciclopedias Sobre El Académico*. Accessed: 01 May 2012. <http://www.esacademic.com/>.

Baobab Tree (*Adansonia digitata*)

Hankey, Andrew. "Adansonia Digitata." *PlantZAfrica.com*. Feb. 2004. (accessed January, 2012).

Hines, Deborah A., and Karlyn Eckman. *Indigenous Multipurpose Trees of Tanzania: Uses and Economic Benefits for People*. Canada: Cultural Survival Canada, 1993.

Sturtevant, E. Lewis. *Sturtevant's Edible Plants of the World*. Ed. U. P. Hedrick. New York: Dover Publications, 1972.

Kapitany, Attila. *Australian Succulent Plants: An Introduction*. Boronia, Victoria, Australia: Kapitany Concepts, 2007.

Soares, Claire. "The tree of life (and its super fruit)." *The Independent*, 17 July 2008.

Argan Tree (*Argania spinosa*)

Axtell, B. L., and R. M. Fairman. *Minor Oil Crops*. Rome: Food and Agriculture Organization of the United Nations, 1992.

Morton, J.F. and G.L. Voss. "The argan tree (*Argania sideroxylon*, Sapotaceae), a desert source of edible oil." *Economic Botany*. 41: 221-233, 1987.

Sturtevant, E. Lewis. *Sturtevant's Edible Plants of the World*. Ed. U. P. Hedrick. New York: Dover Publications, 1972.

Nerd, A., Eteshola, E., Borowy, N., and Mizrahi, Y. "Growth and oil production of argan in the Negev Desert of Israel." *Industrial Crops and Products*. 2: 89-95, 1994.

Prendergast, H.D.V. and Walker, C.C. "The argan, multipurpose tree of Morocco." *Kew Magazine* 9: 76-85, 1992.

Farines M, Soulier J, Charrouf, Soulier R. "Study of the Seed Oil from *Argania spinosa* (L.) Sapotaceae. I. The glyceride fraction." *Revue Francaise des Corps Gras*. 31 (7/8): 283-286, 1984.

"Argania Spinosa." *UA Campus Arboretum*. University of Arizona. Accessed: January 2012. <http://arboretum.arizona.edu/>.

Orwa C, Mutua A , Kindt R , Jamnadass R, Simons A. Agroforestree Database: a tree reference and selection guide version 4.0. 2009. Accessed: January 2012 <http://www.worldagroforestry.org/>.

Purple Hook Berry (*Artabotrys brachypetalus*)

Jolad, S. D.; Hoffmann, J. J.; Schram, K. H.; Cole, J. R.; Tempesta, M. S.; Kriek, G. R.; Bates, R. B. "Uvaricin, a new antitumor agent from *Uvaria accuminata* (Annonaceae)". *The Journal of Organic Chemistry* 47 (16): 3151–3153, 1982.

Champy, Pierre; Hoeglinger, Guenter U.; Feger, Jean; Gleye, Christophe; Hocquemiller, Reynald; Laurens, Alain; Guerineau, Vincent; Laprevote, Olivier; Medja, Fadia; Lombes, Anne; Michel, Patrick P.; Lannuzel, Annie; Hirsch, Etienne C.; Ruberg, Merle. "Annonacin, a lipophilic inhibitor of mitochondrial complex I, induces nigral and striatal neurodegeneration in rats: Possible relevance for atypical parkinsonism in Guadeloupe". *Journal of Neurochemistry* 88 (1): 63–69, 2004.

Odebode, A. C., S. A. Jonker, C. C. Joseph, and S. W. Wachira. "Anti-Fungal Activities of Constituents from *Uvaria scheffleri* and *Artabotrys brachypetalus*." *Journal of Agricultural Sciences* 51.1 79-86, 2006.

Flora of Zimbabwe: Home Page. Accessed: 10 April 2011. <http://www.zimbabweflora.co.zw/>.

Silverhill Seeds. Accessed: 10 April 2011. <http://www.silverhillseeds.co.za/>.

Desert Date (*Balanites aegyptiaca*)

Booth, F. E. M., and G. E. Wickens. *Non-timber Uses of Selected Arid Zone Trees and Shrubs in Africa.* Rome: Food and Agriculture Organization of the United Nations, 1988.

National Research Council. *Lost Crops of Africa: Volume III: Fruits.* Washington, DC: National Academies Press, 2008.

Hines, Deborah A., and Karlyn Eckman. *Indigenous Multipurpose Trees of Tanzania: Uses and Economic Benefits for People.* Canada: Cultural Survival Canada, 1993.

Velvet Sweetberry (*Bridelia mollis*)

Flora of Zimbabwe: Home Page. Accessed: 15 April 2011. <http://www.zimbabweflora.co.zw/>.

Motlhanka, D.M.T., P. Motlhanka, and T. Selebatso. "Edible Indigenous Wild Fruit Plants of Eastern Botswana." *International Journal of Poultry Science* 7.5 (2008): 457-60.

Num Berry (*Carissa bispinosa*)

Coates Palgrave, M. 2002. *Keith Coates Palgrave Trees of southern Africa*, edn 3. Struik, Cape Town.

Codd, L.E. 1963. Apocynaceae. *Flora of southern Africa* 26: 250-258. Botanical Research Institute, Pretoria.

Glen, H.F. 2002. *Cultivated plants of southern Africa.* Jacana, Johannesburg.

Leeuwenberg, A.J.M. & Van Dilst, F.J.H. 2001. *Series of revisions of Apocynaceae XLIX Carissa L.* Wageningen University Papers, Blackhuys Publishers, Leiden, The Netherlands.

Palmer, E. & Pitman, N. 1972. *Trees of southern Africa*, vol. 3. Balkema, Cape Town.

Swart, W.J. 1982. Survival off the bush-5: The yummy num-nums. *Farmer's Weekly*, October 1982: 58, 59.

Van Wyk, B.-E. & Gericke, N. 2000. *People's plants: a guide to useful plants of southern Africa.* Briza, Pretoria.

Khat (*Catha edulis*)

Hoffer, Abram, and Humphry Osmond. *The Hallucinogens.* New York: Academic, 1967.

Pooley, E. 1993. *The complete field guide to trees of Natal, Zululand and Transkei.* Natal Flora Publications Trust, Durban.

Schmidt, E., Lötter, M. & McCleland, W. 2002. *Trees and shrubs of Mpumalanga and Kruger National Park.* Jacana, Johannesburg.

Van Wyk, B.-E., Van Oudtshoorn, B. & Gericke, N. 1997. *Medicinal plants of southern Africa.* Briza Publications, Pretoria.

Van Wyk, B. & Van Wyk, P. 1997. *Field guide to trees of southern Africa.* Struik, Cape Town.

Tick Berry (*Chrysanthemoides monilifera subsp. monilifera*)

Hutchings, A. 1996. *Zulu Medicinal Plants, An Inventory.* University of Natal Press.

Palgrave, K.C. 1977. *Trees of Southern Africa.*

Van Breda, P.A., & Barnard, S.A. 1992. *100 Veld plants of the winter rainfall region.* Department of Agricultural Development, Bull. 422.

Kei Apple (*Dovyalis caffra*)

Lost Crops of Africa. Washington: National Academies, 2008.

Morton, Julia F. *Fruits of Warm Climates.* Miami, FL, 1987.

Cape Ebony (*Euclea pseudebenus*)

Fourie, Tobias. "*Euclea pseudebenus*, Cape Ebony, Wild Ebony." *Gateway Africa.* Accessed: 17 Apr. 2011. <http://www.gateway-africa.com/>.

"*Euclea pseudebenus* E.Mey. Ex A.DC." *Kyffhäuser Plants.* Accessed: 23 Feb. 2012. <http://www.kyffhauser.co.za/>.

"Silverhill Seeds: Seeds from Africa." *Silverhill Seeds: Seeds from Africa.* Accessed: 17 Apr. 2011. <http://www.silverhillseeds.co.za/>.

"Peace Parks Foundation: The Global Solution." *Peace Parks Foundation.* Accessed: 23 Feb. 2012. <http://www.peaceparks.org/>.

Fiori (*Grewia tenax*)

Mohammed Elhassan, G.O., and S.M. Yagi. "Nutritional Composition of *Grewia* Species (*Grewia tenax* (Forsk.) Fiori, *G. Flavescens* Juss and *G. Villosa* Willd) Fruits." *Advance Journal of Food Science and Technology* 2.3 (2010): 159-62.

Al-Said, Mansour S., Ramzi A. Mothana, Mohammed O. Al-Sohaibani, and Syed Rafatullah. "Ameliorative Effect of *Grewia tenax* (Forssk) Fiori Fruit Extract on CCl4-Induced Oxidative Stress and Hepatotoxicity in Rats." *Journal of Food Science* 76.9 (2011): T200-206.

Baumer M. 1983. *Notes on trees and shrubs in arid and semi-arid regions.* Rome FAO. Forestry Division.

Beentje HJ. 1994. *Kenya trees, shrubs and lianas.* National Museums of Kenya.

IBPGR-Kew. 1984. *Forage and browse plants for arid and semi-arid Africa.* Rome. IBPGR.

Vogt K. 1995. *A field guide to the identification, propagation and uses of common trees and shrubs of dryland Sudan.* SOS Sahel International (UK).

von Maydell HJ. 1986. *Trees and shrubs of the Sahel - their characteristics and uses.* GTZ 6MBH, Eschborn.

Zabala (*Lardizabala biternata*)

Marticorena, Alicia. *Plantas Trepadoras, Epífitas Y Parásitas Nativas De Chile: Guía De Campo.* Concepción, Chile: CORMA, 2010.

"Description and Images of *Lardizabala biternata.*" *Chileflora.* Accessed: 17 Mar. 2012. <http://www.chileflora.com/>.

"*Lardizabala biternata.*" *Chilebosque.* Accessed: 17 Mar. 2012. <http://www.chilebosque.cl/>.

Plants for a Future Database. www.pfaf.org (accessed March, 2012).

Goji Berry (*Lycium barbarum*)

"Goji Berry Amount and Nutrition Facts Conversion." *Convert-to.com.* Accessed: 10 June 2012. <http://convert-to.com/>.

Yunyun, Liu. "Dry No More." *Beijing Review*, 16 Oct. 2008. Accessed: 21 Mar. 2012. <http://www.bjreview.com.cn/>.

"Ningxia Climate & Weather." *China Maps.Info*, 2007. Accessed: 21 Mar. 2012. <http://www.chinamaps.info/>.

Plants for a Future Database. www.pfaf.org (accessed March, 2012).

Skae, Teya. "Super Goji Berry: Choosing the Best Source Among the Marketing Hype." Natural News. *Natural News Network*, 14 Feb. 2008. Accessed: 21 Mar. 2012. <http://www.naturalnews.com/>.

Reeve, Vivienne E., Munif Allanson, Sondur Jayappa Arun, Diane Domanski, and Nicole Painter. "Mice Drinking Goji Berry Juice (*Lycium barbarum*) Are Protected from UV Radiation-induced Skin Damage via Antioxidant Pathways." *Photochemical & Photobiological Sciences* 9.4 (2010).

Tang, Wai-Man, Enoch Chan, Ching-Yee Kwok, Yee-Ki Lee, Jian-Hong Wu, Chun-Wai Wan, Robbie Chan, Peter Yu, and Shun-Wan Chan. "A Review of the Anticancer and Immunomodulatory Effects of *Lycium barbarum* Fruit." *Inflammopharmacology* (2011): 1-8.

Moringa (*Moringa oleifera*)

Sturtevant, E. Lewis. *Sturtevant's Edible Plants of the World.* Ed. U. P. Hedrick. New York: Dover Publications, 1972.

Janick, Jules, and Robert E. Paull. *The Encyclopedia of Fruit & Nuts.* Wallingford, UK: CABI North American Office, 2008.

"Moringa oleifera." *World Agroforestry Centre.* Accessed: 25 Mar. 2012. <http://www.worldagroforestrycentre.org/>.

"African Moringa." *Echo Plant Information Sheet.* ECHO, Inc, 2006.

Doerr, Beth. "Moringa Water Treatment." *An Echo Technical Note.* ECHO, Inc, 2005.

"Moringa." *Echo Plant Information Sheet.* ECHO, Inc, 2006.

Rosa, K.R. Dalla. "Moringa oleifera: A Perfect Tree for Home Gardens." *Winrock International.* Forest, Farm, and Community Tree Network, Apr. 1993. Accessed: 25 Mar. 2012. <http://www.winrock.org/>.

Lucumillo (*Myrcianthes coquimbensis*)

Saldías, Gabriela. "Propagación de *Myrcianthes coquimbensis* (Myrtaceae), especie endémica de Chile con alto valor ambiental para su uso en proyectos de paisaje." *Chloris Chilensis: Revista Chilena De Flora Y Vegetacion.* Accessed: 25 Mar. 2012. <http://www.chlorischile.cl/>.

García-Guzman, Patricio, Andrea P. Loayza, Danny E. Carvajal, Luis Letelier, and Francisco A. Squeo. "The Ecology, Distribution and Conservation Status of *Myrcianthes coquimbensis*: A Globally Endangered Endemic Shrub of the Chilean Coastal Desert." *Plant Ecology & Diversity.*

"Description and Images of *Myrcianthes Coquimbensis*." *Chileflora.com*. Accessed: 25 Mar. 2012. <http://www.chileflora.com/>.

Landrum, Leslie R., and Francesca T. Grifo. "*Myrcianthes* (Myrtaceae) in Chile." *Brittonia* 40.3 (1988): 290-93.

Myrtle (*Myrtus communis*)

Aydin, C., and M. Ozcan. "Determination of Nutritional and Physical Properties of Myrtle (*Myrtus communis* L.) Fruits Growing Wild in Turkey." *Journal of Food Engineering* 79.2 (2007): 453-58.

A Guide to Medicinal Plants in North Africa. Málaga: IUCN Centre for Mediterranean Cooperation, 2005.

Plants for a Future Database. www.pfaf.org (accessed April, 2011).

Transvaal Kooboo-Berry (*Mystroxylon aethiopicum subsp. aethiopicum*)

Coates Palgrave, M. 2002. *Keith Coates Palgrave Trees of southern Africa*, edn 3. Struik, Cape Town.

Schmidt, E., Lötter, M. & McCleland, W. 2002. *Trees and shrubs of Mpumalanga and Kruger National Park*. Jacana, Johannesburg.

Van Wyk, B. & Van Wyk, P. 1997. *Field guide to trees of southern Africa*. Struik, Cape Town.

Venter, F. & Venter, J.A. 1996. *Making the most of indigenous trees*. Briza Publications, Pretoria.

Tortoise Berry (*Nylandtia spinosa*)

Viljoen, Cherise. "*Nylandtia spinosa*." *PlantZAfrica.com*. Kirstenbosch National Botanical Garden, Sept. 2005. Accessed: 17 Apr. 2011. <http://www.plantzafrica.com/>.

Amabeoku, G. "Anticonvulsant Activity of *Nylandtia spinosa* L. Dumont (Polygalaceae) Aqueous and Methanol Leaf Extracts in Mice." *Human & Experimental Toxicology* 27.11 (2008): 811-18.

Silverhill Seeds. Accessed: 28 March 2012. <http://www.silverhillseeds.co.za/>.

"Nylandtia Spinosa Herba." *PlantZAfrica.com*. Accessed: 29 Mar. 2012.

Sacred Succulents. Accessed: 28 March 2012. <http://sacredsucculents.com/>.

Markouba Grass (*Panicum turgidum*)

Anthelme, Fabien, and Richard Michalet. "Grass-to-tree Facilitation in an Arid Grazed Environment (Aïr Mountains, Sahara)." *Basic and Applied Ecology* 10.5 (2009): 437-46.

"Edible Perennial Plants Native to Hot Deserts - the Solution to the Invading Sand Desert." *Eden Foundation*. Accessed: 01 Feb. 2012. <http://www.eden-foundation.org/>.

"Panicum turgidum." *FAO: FAO Home*. Food and Agriculture Organization of the United Nations. Accessed: 01 Feb. 2012. <http://www.fao.org/ag/>.

Jacket Plum (*Pappea capensis*)

Sobey, Roxanne, and Helen Watson. "An Investigation into the Biofuel Potential of Two Trees Indigenous to Southern Africa." *Compete* 2 (2008): 12-14.

Silverhill Seeds. Accessed: 10 April 2011. <http://www.silverhillseeds.co.za/>.

Coates Palgrave, M. 2002. *Keith Coates Palgrave Trees of southern Africa*, edn 3. Struik, Cape Town

Hutchings, A. 1996. *Zulu medicinal plants: an inventory*. University of Natal Press, Pietermaritzburg

Palmer, E. & Pitman, N. 1972. *Trees of southern Africa, vol. 2*. Balkema, Cape Town.

Pooley, E. 1993. *The complete field guide to trees of Natal, Zululand and Transkei*. Natal Flora Publications Trust, Durban.

Schmidt, E., Lötter, M. & McCleland, W. 2002. *Trees and shrubs of Mpumalanga and Kruger National Park*. Jacana, Johannesburg.

Venter, S. & Venter, J. 1985. *Making the most of indigenous trees*. Briza Publications, Pretoria.

Galupa (*Passiflora pinnatistipula*)

"Description and Images of Passiflora pinnatistipula." *Chileflora*. Accessed: 20 Apr. 2011. <http://www.chileflora.com/>.

Marticorena, Alicia. *Plantas Trepadoras, Epífitas Y Parásitas Nativas De Chile: Guía De Campo*. Concepción, Chile: CORMA, 2010.

Møller Jørgensen, Peter, and Roberto Vásquez. "A Revision of *Passiflora* Sections *Insignes* and ×*Inkea* (Passifloraceae)." *Anales Del Jardín Botánico De Madrid* 66.1 (2009): 35-53.

"*Passiflora pinnatistipula*." *Chilebosque*. Accessed: 17 Mar. 2012. <http://www.chilebosque.cl/>.

Single-Leaf Pinyon (*Pinus monophylla*)

Ronald M. Lanner, 1981. *The Piñon Pine: A Natural and Cultural History*. University of Nevada Press.

Gerry Moore, Bruce Kershner, Craig Tufts, Daniel Mathews, Gil Nelson, Richard Spellenberg, John W. Thieret, Terry Purinton and Andrew Block (2008). *National Wildlife Federation Field Guide to Trees of North America*. New York: Sterling.

Zouhar, Kristin L. 2001. *Pinus monophylla*. In: Fire Effects Information System, [Online]. U.S. Department of Agriculture, Forest Service, Rocky Mountain Research Station, Fire Sciences Laboratory (Producer). Accessed: 10 Apr. 2012. <http://www.fs.fed.us/database/feis/>.

Geisler, Malinda. "Pine Nuts Profile." *Agricultural Marketing Resource Center*. Iowa State University, Apr. 2011. Accessed: 10 Apr. 2012. <http://www.agmrc.org/>.

Stone Pine (*Pinus pinea*)

Loewe, Verónica M., Claudia R. Delard, and Alejandro G. Venegas. "Pine Nut (*Pinus pinea* L.) Production, an Alternative for Temperate Areas." *Asia Pacific Agroforestry Newsletter* 39 (Dec. 2011): 4-7.

Trap, Louis. "*Pinus pinea*: An Edible Nut Pine of Many Uses." *The Australian New Crops Newsletter* 6 (July 1996).

Gilman, Edward F., and Dennis G. Watson. "*Pinus pinea* Stone Pine." *Southern Group of State Foresters*. US Department of Agriculture Forest Service, Oct. 1994. Accessed: Apr. 2012. <http://hort.ufl.edu/>.

Plants for a Future Database. www.pfaf.org (accessed April, 2012).

"Infected Trees with *Tuber borchii*." *Micofora.com*. Micologia Forestal & Aplicada. Accessed: 11 Apr. 2012. <http://www.micofora.com/>.

"Bianchetto Truffle (*Tuber borchii*)." Truffles & Mushrooms Ltd, 1 Nov. 2010. Accessed: 11 Apr. 2012. <http://www.trufflesandmushrooms.co.nz/>.

Lucuma (*Pouteria lucuma*)

Lost Crops of the Incas: Little-known Plants of the Andes with Promise for Worldwide Cultivation. Washington, D.C.: National Academy, 1989.

Morton, Julia F. *Fruits of Warm Climates.* Miami, FL, 1987.

"Description and Images of *Pouteria splendens* (Lucumo Silvestre)" *ChileFlora.* Accessed: 13 Apr. 2012. <http://www.chileflora.com/>.

"Lucuma." *Family Content.* 1 Jan. 2008. Accessed: 13 Apr. 2012. <http://www.family-content.com/>.

Pinto, M.D-S., Ranilla, L.G., Apostolidis, E., Lajolo, F.M., Genovese, M.I. and Shetty, K. (2009). "Evaluation of anti-hyperglycemia and anti-hypertension potential of native Peruvian fruits using *in vitro* models." *J. Medicinal Food*, 12: 278-291.

Rojo, Leonel E., Caren M. Villano, Gili Joseph, Barbara Schmidt, Vladimir Shulaev, Joel L. Shuman, Mary Ann Lila, and Ilya Raskin. "Wound-healing Properties of Nut Oil from *Pouteria Lucuma.*" *Journal of Cosmetic Dermatology* 9.3 (2010): 185-95.

Desert Quandong (*Santalum acuminatum*)

Watling, Jennifer R., and Ben Lethbridge. *Water and Nutrient Relations of Parasitic Quandong and Acacia Host Species: A Report for the Rural Industries Research and Development Corporation.* Barton, A.C.T.: Rural Industries Research and Development, 2007.

Tennakoon K.U., Pate J.S. & Arthur D. 1997. Ecophysiological aspects of the woody root hemiparasite *Santalum acuminatum* (R. Br.) A. DC. and its common hosts in South Western Australia. *Annals of Botany*, 80: 245-56.

Loveys B. R., Tyerman S. D., Loveys B. R. 2001. Transfer of photosynthate and naturally occurring insecticidal compounds from host plants to the root hemiparasite *Santalum acuminatum* (Santalaceae). *Australian Journal of Botany* 49: 9–16.

Smith, Richard M., and Pamela A. Smith. "An Assessment of the Composition and Nutrient Content of an Australian Aboriginal Hunter-gatherer Diet." *Australian Aboriginal Studies* (2003/2): 39-52.

"Quandong Fruit Flesh." *Food Standards Australia New Zealand.* Accessed: 24 Apr. 2012. <http://www.foodstandards.gov.au/>.

"*Santalum acuminatum.*" *Ecocrop.* Food and Agriculture Organization of the UN. 24 Apr. 2012. <http://ecocrop.fao.org/>.

"Fact Sheet: Quandong Garden." *Gardening Australia.* ABC. Accessed: 24 Apr. 2012. <http://www.abc.net.au/>.

"*Santalum acuminatum.*" *Australian National Botanic Gardens.* Australian Government. Accessed: 24 Apr. 2012. <http://www.anbg.gov.au/>.

"*Santalum acuminatum.*" *Australian Native Plants Society (Australia).* Accessed: 24 Apr. 2012. <http://anpsa.org.au/>.

"Factsheet - *Santalum acuminatum.*" *Florabank.* Accessed: 24 Apr. 2012. <http://www.florabank.org.au/>.

"Factsheet - *Santalum spicatum.*" *Florabank.* Accessed: 24 Apr. 2012. <http://www.florabank.org.au/>.

Plants for a Future Database. www.pfaf.org (accessed April, 2012).

Mongongo (*Schinziophyton rautanenii*)

Booth, F. E. M., and G. E. Wickens. *Non-timber Uses of Selected Arid Zone Trees and Shrubs in Africa.* Rome: Food and Agriculture Organization of the United Nations, 1988.

Nerd, A., J.A. Aronson, and Y. Mizrahi. 1990. Introduction and domestication of rare and wild fruit and nut trees for desert areas. p. 355-363. In: J. Janick and J.E. Simon (eds.), Advances in new crops. Timber Press, Portland, OR.

Graz, F. Patrick. "The Growth of Schinziophyton rautanenii Seedlings under Different Shade Conditions." *Dinteria* 28 (2003): 44-46.

Marula (*Sclerocarya birrea subsp. caffra*)

Lost Crops of Africa. Washington: National Academies, 2008.

Nerd, A. and Y. Mizrahi. 1993. Domestication and introduction of marula (*Sclerocarya birrea subsp. caffra*) as a new crop for the Negev desert of Israel. p. 496-499. In: J. Janick and J.E. Simon (eds.), New crops. Wiley, New York.

Zharare, P., and N. Dhlamini. "Characterization of Marula (*Sclerocarya caffra*) Kernel Oil and Assessment of Its Potential Use in Zimbabwe." *The Journal of Food Technology in Africa* 5.4 (2000): 126-28.

"*Sclerocarya birrea subsp. caffra.*" *Survey of Economic Plants for Arid and Semi-Arid Lands (SEPASAL) Database.* Kew Royal Botanic Gardens. Accessed: 29 Apr. 2012. <http://www.kew.org/>.

"*Sclerocarya birrea ssp. caffra.*" *AgroForestryTree Database.* World Agroforestry Centre. Accessed: 3 Apr. 2011. <http://www.worldagroforestrycentre.org/>.

Van Staden, J. M. "*Sclerocarya birrea ssp. caffra.*" *EcoPort.* ARC - Agricultural Research Council Pretoria, South Africa. Accessed: 29 Apr. 2012. <http://ecoport.org/>.

"*Sclerocarya birrea ssp. caffra* (Marula)." *Skyfield Tropical: Encyclopedia: Rare Fruit Trees.* Accessed: 3 Apr. 2011. <http://www.skyfieldtropical.com/>.

Monkey Orange (*Strychnos cocculoides*)

"*Strychnos cocculoides.*" *The Culture Sheet.* Southampton Centre for Underutilised Crops. Accessed: 03 Apr. 2011. <http://culturesheet.org/loganiaceae:strychnos:cocculoides>.

Motlhanka, D.M.T., P. Motlhanka, and T. Selebatso. "Edible Indigenous Wild Fruit Plants of Eastern Botswana." *International Journal of Poultry Science* 7.5 (2008): 457-60.

Glen, Hugh. "*Strychnos cocculoides.*" *PlantZAfrica.com.* KwaZulu-Natal Herbarium. Accessed: 05 May 2012. <http://www.plantzafrica.com/>.

"*Strychnos cocculoides.*" *Ecocrop.* FAO - Food and Agriculture Organization of the UN. Accessed: 05 May 2012. <http://ecocrop.fao.org/>.

Jujube (*Ziziphus jujuba*)

Plants for a Future Database. www.pfaf.org (accessed May, 2012).

"Jujube." Fruit Facts. *California Rare Fruit Growers.* Accessed: 3 April 2011. <http://www.crfg.org/>.

Ayers, R. S., and D. W. Westcot. *Water Quality for Agriculture.* Rome: Food and Agriculture Organization of the United Nations, 1985.

Mengjun, L. 2003. Genetic Diversity of Chinese Jujube (*Ziziphus jujube* Mill.). *Acta Hort.* (ISHS) 623: 351-355.

Zhang, Ping, You-Ke Wang, Jing-Wu Zhan, Xing Wang, and Peng Wu. "Scheduling Irrigation for Jujube (*Ziziphus jujuba* Mill.)." *African Journal of Biotechnology* 9.35 (2010): 5694-703.

Shangguan, Z. P., and S. X. Zheng. "Ecological Properties of Soil Water and Effects on Forest Vegetation in the Loess Plateau." *International Journal of Sustainable Development & World Ecology* 13.4 (2006): 307-14.

Christ's Thorn Jujube (*Ziziphus spina-christi*)

"Nutritional Study on Ziziphus spina-christi". *Eden Foundation*. Accessed: 15 May 2012. <http://www.eden-foundation.org/>.

Baumer, Michel. *Notes on Trees and Shrubs in Arid and Semi-arid Regions*. Rome: Food and Agriculture Organization of the United Nations, 1983.

Global Forestry Compendium. CD-ROM. CABI. 2000.

Firewood Crops: Shrub and Tree Species for Energy Production. Washington D.C: National Academy of Sciences, 1980.

Sohail, M., A. S. Saied, J. Gebauer, and A. Buerkert. "Effect of NaCl Salinity on Growth and Mineral Composition of *Ziziphus spina-christi* (L.) Willd." *Journal of Agriculture and Rural Development in the Tropics and Subtropics* 110.2 (2009): 107-114.

Motlhanka, D.M.T., P. Motlhanka, and T. Selebatso. "Edible Indigenous Wild Fruit Plants of Eastern Botswana." *International Journal of Poultry Science* 7.5 (2008): 457-60.

Pandey, Anjula, Rakesh Singh, J. Radhamani, and D. C. Bhandari. "Exploring the potential of *Ziziphus nummularia* (Burm. F.) Wight Et Arn. from drier regions of India." *Genetic Resources and Crop Evolution* 57.6 (2010): 929-36.

Gorai, Mustapha, Mouna Maraghni, Mohamed Neffati. "Relationship between phenological traits and water potential patterns of the wild jujube *Ziziphus lotus* (L.) Lam. in southern Tunisia". *Plant Ecology & Diversity* 3.3 (2010): 273-280.

About the Author: A native of the sun-soaked town of Santa Barbara, California, Scott O'Bar gained a passion for vegetable gardening at an early age. He obtained an AS degree in Environmental Horticulture from Santa Barbara City College and has worked at various horticultural jobs, including a position at one of Santa Barbara's world famous botanic gardens: Lotusland. He is certified in permaculture design, a practice which led him beyond the mainstream methods of food production.

Scott's concern over the mismanagement of the world's freshwater, has led him to scour the botanical world for key edible species that require little irrigation compared to today's major crop species. He is a visionary proponent of food forestry, and a self-described rare & edible plant enthusiast who has spent years researching alternative systems of food production, which require little energy and few external inputs.

Scott hopes to foster more firsthand research involving drought-tolerant edible plants, and seeks to ease the burden of drought-afflicted regions throughout the world by installing multi-species agroforests that simultaneously heal the land and produce food with minimal irrigation.